检测基础
Fundamentals of Testing
（双语）

主编　张保林
参编　张　博　弋　楠　杨兵兵　宋丽平
主审　李红莉

北京理工大学出版社
BEIJING INSTITUTE OF TECHNOLOGY PRESS

内 容 简 介

本教材是中国特色高水平高职学校和专业建设计划教材，根据《国家职业教育改革实施方案》及《高等职业学校专业教学标准（试行）》编写而成。教材共分为 16 章，主要介绍了金属的力学性能测试、金相检测，热处理分析、金属焊接性分析、超声检测、射线检测、磁粉检测、渗透检测、涡流检测、其他无损检测方法，以及质量控制等内容。本教材内容涵盖了理化测试方法、无损检测方法，以及与检测密切相关的基础中英文专业知识，具有将检测基础专业知识及科技英语紧密结合的双语教材的特点。

本教材可作为职教本科及高职高专院校材料类专业、金属材料检测类专业的双语教材，也可作为理化测试、无损检测与热处理技术双语人员的培训教材。

图书在版编目（CIP）数据

检测基础 = Fundamentals of Testing：英、汉 / 张保林主编. -- 北京：北京理工大学出版社，2024.5
ISBN 978-7-5763-3928-4

Ⅰ.①检…　Ⅱ.①张…　Ⅲ.①测试技术-英、汉　Ⅳ.①TB4

中国国家版本馆 CIP 数据核字（2024）第 090868 号

责任编辑：王梦春		**文案编辑**：辛丽莉	
责任校对：周瑞红		**责任印制**：李志强	

出版发行 / 北京理工大学出版社有限责任公司
社　　址 / 北京市丰台区四合庄路 6 号
邮　　编 / 100070
电　　话 / （010）68914026（教材售后服务热线）
　　　　　　（010）68944437（课件资源服务热线）
网　　址 / http：//www.bitpress.com.cn

版 印 次 / 2024 年 5 月第 1 版第 1 次印刷
印　　刷 / 涿州市新华印刷有限公司
开　　本 / 787 mm×1092 mm　1/16
印　　张 / 11.5
字　　数 / 188 千字
定　　价 / 69.80 元

前　言

　　《检测基础》双语教材是依据高等职业技术学院理化测试与质检技术专业人才的培养需要而编写的。教材共分为16章，主要介绍了金属的力学性能测试、金相检测、热处理分析、金属焊接性分析、超声检测、射线检测、磁粉检测、渗透检测、涡流检测、其他无损检测方法，以及质量控制等内容。本教材涵盖了理化测试方法、无损检测方法，以及与检测密切相关的中英文专业基础知识，具有检测基础专业知识与科技英语紧密结合的双语教材特点。

　　《检测基础》双语教材是理化测试与质检技术专业进行岗位能力培养的一门专业拓展课程适用教材。该教材针对理化测试与无损检测专业人才需要掌握的专业科技英语组织教学内容，按照双语学习的特点设计教学环节，为检测岗位专业能力需求提供科技英语读写职业能力，为培养检测专业技术技能型人才提供保障。

　　编写《检测基础》双语教材的目的是让理化测试与无损检测技术应用人才掌握从事岗位需要的专业英语词汇和基础知识，提高阅读、理解本专业科技英语的能力，具有阅读专业英文资料和翻译科技英文的能力。本教材还适合于智能焊接技术等相关专业学生及工程技术人员使用。

　　本教材由陕西工业职业技术学院张保林副教授负责编写。具体编写分工如下：张保林编写第1、2、3、7、9、11、12、14、15、16章，陕西工业职业技术学院弋楠副教授编写第10、13章，陕西工业职业技术学院张博教授编写第4、5章，陕西工业职业技术学院杨兵兵教授编写第8章，陕西工业职业技术学院宋丽平副教授编写第6章，张保林进行了统稿。本教材由陕西工业职业技术学院材料工程学院理化测试与质检技术专业带头人李红莉副教授审阅。

　　在编写过程中，编者参阅了国内外出版的有关教材和资料，同时也得到了陕西省技能大师、国核宝钛锆业股份公司高级工程师卢辉的帮助。在此，编者表示衷心的感谢！

　　由于编者水平有限，书中不妥之处在所难免，恳请读者批评指正。

编　者

Contents
目　录

Chapter 1　Properties of Metals

Let us see why metals have come to play so large a part in man's activities. Wood and stone are both older in use, yet to a considerable extent they have been supplanted by the metals. The cause of the increasing use of metals is to be found in their characteristic properties, such as: strength, or ability to support mass without bending or breaking; toughness, or ability to bend rather than break under a sudden blow; resistance to atmospheric destruction; and malleability, or ability to be formed into the desired shape. Malleability of a metal is also known as the ability to deform permanently under compression without rupture. It is the property which allows the hammering and rolling of metals into thin sheets.

Metals can be cast into varied and intricate shapes weighing from a few ounces (ounce is a unit of mass, the symbol is oz, 1 oz = 28,3495g) to many tons. Their plasticity, or ability to deform without rupture, makes them safe to use in all types of structures, and also allows their formation into required shapes through forging and other operations. Metals also possess the important property of being weldable. Of all the engineering materials only metals are truly weldable and repairable.[1] Other materials used in engineering constructions, including glass, stone, and wood, usually are destroyed when the structure is no longer usable. An unusable bridge, ship, or boiler made of metal usually is cut into easily handled sections, put in a furnace, remelted, cast, and finally worked in the making of a new ship, bridge, or boiler.[2]

All of this represents a remarkable combination of properties, one possessed by no other class of materials. Some metals also possess additional, special properties. Two of which are power to conduct electric current and the ability to be magnetized. The selection of the proper metal or alloy for a given use is an important part of the practice of metallurgy. Because iron and steel are used in larger quantities than any of the other metals, it is common practice to divide metallurgical materials into ferrous, or iron-bearing, and nonferrous, or those containing no iron, or only small proportions of iron.

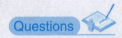

1. Why were wood and stone supplanted by metals?

2. What light mass metals do you know?

3. What is the property allowing forging and rolling metals?

4. Can wood and stone be used for conducting electric current?

5. Why is metal considered more repairable than stone and wood?

①Of all the engineering materials only metals are truly weldable and repairable.

在所有的工程材料中，只有金属材料才真正地具有可焊接性和可修补性。

句中 of 表示范围，译为"在……当中"。

②An unusable bridge, ship, or boiler made of metal usually is cut into easily handled sections, put in a furnace, remelted, cast, and finally worked in the making of a new ship, bridge, or boiler.

由金属制成的废弃桥梁、船只、锅炉通常被切割成易处理的小块，投入炉中，然后进行熔化、铸造，最后制造出新的船只、桥梁、锅炉。

本句中 put、remelted、cast、worked 和 cut 一样都是过去分词，它们构成并列谓语。

金属的性能

让我们来看看为什么金属在人类生活中起着如此重要的作用。在工程应用上，木材和石头是两个用得比较早的工程材料，然而，在很大程度上，已经被金属替代。增加金属材料用途的原因是它们的性能好，如强度，能够承受质量而不弯曲和折断；韧性，即在突然的冲击下弯曲而没有断裂；耐大气破坏性；可锻性是指能够成形所需要的形状。金属的可锻性也称为在受压条件下没有断裂的永久变形能力。它是允许金属被锤锻和轧制成薄板的性能。

金属能够铸造成各种各样复杂的形状，质量可以从几盎司到许多吨。它们

的塑性，即变形而不断裂的能力使得金属能够安全地用于所有的工程结构类型，并且可以通过锻造或其他操作处理使金属变形成所需要的形状。金属也具有可以焊接的重要性能，在所有的工程材料中，只有金属材料才能真正具有焊接性和可修补性能。其他用于工程结构上的材料包括玻璃、石头、木材，当这个结构不再使用时，通常就被破坏，而不能再用。由金属材料制造的工程结构，如不能够再使用的旧的桥梁、轮船或锅炉，通常可以切割成容易处理的小块，送入熔炼炉中，重新进行熔化、铸造，并且最终加工成新的轮船、桥梁或锅炉。

所有这些，表示金属材料的一个显著的性能组合，这是别的种类的材料所没有的。一些金属也具有附加的、特殊的性能，即传导电流和具有磁性两种性能。对于给定的用途，选择合适的金属或合金是实际冶金的一个重要部分。因为铁和钢比其他任何金属的用量都大，通常的做法是将冶金材料分成铁质金属和非铁质金属，即不含铁或仅有少量铁的金属。

Reading Material

Hooke's Law

Some solid bodies have a tendency to maintain their shape, and when deformed by an external force, they return to their original shape as soon as the force is removed. This property is known as elasticity. A fundamental law in this field, Hooke's Low of elasticity, states that the deformation of a solid body is proportional to the force acting on it, provided the force does not exceed a certain limit.

It will be easier to deal with Hooke's Law quantitatively if we pause to define a pair of technical terms, "stress" and "strain". Stress refers to the internal forces created within a material as a result of forces applied to it. It is F/A, the applied force divided by the cross section of material that opposes the force.

Strain is a measure of how much a body is deformed by a stress. The basic types of deformation are stretching, bending, and twisting. Let us look at the lengthening of a stressed wire. The strain of the wire is the elongation divided by the length.

If we hang a weight on a piece of wire 1-meter long, the wire will stretch a certain amount. The same mass hung on a piece of the same wire 2-meter long will cause it to stretch twice as much, since each meter will elongate exactly the same as it did before. In both cases the strain is the same.

We can restate Hooke's Law in a more useful way by saying stress is proportional to strain，or stress/strain＝a constant.

New Words

1. activity *n.* 活跃，活动性

2. considerable *adj.* 相当大的，值得考虑的

3. supplant *vt.* 排挤掉，代替

4. characteristic *adj.* 特有的；*n.* 特性，特征

5. toughness *n.* 韧性

6. bend *v.* 弯曲

7. resistance *n.* 反抗，抵抗，阻力，电阻

8. atmospheric *adj.* 大气的

9. destruction *n.* 破坏，毁灭

10. malleability *n.* 延展性；可锻性

11. deform *v.* （使）变形

12. permanently *adv.* 永存地，不变地

13. compression *n.* 浓缩，压缩

14. rupture *n. & v.* 破裂

15. hammer *n.* 铁锤；*v.* 锤击

16. roll *vt.* 辗，轧

17. cast *n.* 铸件；*v.* 铸造

18. vary *vt.* 改变；*vi.* 变化

19. intricate *adj.* 复杂的

20. ounce *n.* 盎司（质量单位，符号为 oz，1 oz＝28.349 5g）

21. ton *n.* 吨

22. forging *n.* 锻造

23. weldable *adj.* 可焊的

24. boiler *n.* 锅炉

25. furnace *n.* 炉子，熔炉

26. represent *vt.* 表现

27. remarkable *adj.* 值得注意的，显著的

28. metallurgical *adj.* 冶金学的

29. property *n.* 性质，特性

30. magnetize *vt.* 使磁化

Phrases and Expressions

1. to a considerable extent 在相当大的程度上

2. rather than 而不是

3. no longer 不再

4. made of 由……制成

Glossary of Terms

1. characteristic properties 特征性质

2. atmospheric destruction 大气破坏

3. malleability 可锻性

4. intricate shape 复杂形状

5. plasticity 塑性

6. conduct electric 导电性

7. elasticity 弹性

8. cross-section 截面积

9. special property 特殊性能

科技英语语法的特点

　　科技英语是从事科学技术活动时所使用的英语，它主要指描述、探讨自然科学各专业的著作、论文、实验报告、科技实用手段（包括仪器、仪表、机械、工具等）的结构描述和操作说明等。科技英语由于其内容、使用领域和语篇功能的特殊性，也由于科技工作者长期以来的语言使用习惯，形成了自身的一些特点，使其在许多方面有别于日常英语、文学英语等语体。

1. 专业名词、术语多

科技英语专业性强、文体正式，使用大量的专业名词和术语。

2. 第三人称句多

科技英语的一个显著特点就是很少有第一，第二人称的句子，这是由于科技文体的主要目的在于阐述科学事实、科学发现、实验结果等。需要说明的是，在科技英语中有时由于行文的需要，也会使用一些第一、第二人称。但与其他文体相比要少得多。

3. 被动语态多

科技英语的主要目的是表述科学发展观、科学事实、实验报告和各类说明等，以客观陈述为主，而不是以描述活动的完成为主，因此在科技英语中大量使用被动语态。

4. 非谓语动词多

科技英语力求简洁明了，结构紧凑，大量使用非谓语动词包括过去分词、现在分词、动名词和动词不定式。

5. 长句子多

在科技英语中经常使用长句子。这主要是因为在阐述科学事实、科学现象等事物的内在联系、逻辑关系和解释一些科技术语和名词时，需要大量使用各种从句（尤其是定语从句）以及介词短语、形容词短语、分词短语或副词等作后置定语，或者使用分词短语表示伴随情况等。

Chapter 2   Tension Testing

Design of structures and systems requires determination of component dimensions and is based on the appropriate mechanical properties of materials.

Tension Testing   The tension testing is the test most commonly used to evaluate the mechanical properties of materials. A typical load-elongation curve for a pure metal is shown in the figure below.

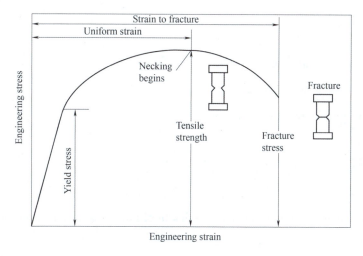

A typical load-elongation curve for a pure metal

A number of important quantities can be calculated from the load-elongation or stress-strain curve of a material, namely:

(1) **Modulus of Elasticity**   This is defined as the tensile stress divided by the tensile strain for elastic deformation and so is the slope of the linear part of the stress-strain curve.

(2) **Yield Strength**   When a material under tension reaches the limit of its elastic strain and begins to flow plastically, it is said to have yielded.[1] The yield strength is then the stress at which plastic flow starts.

(3) **Tensile Strength**   This is defined as the maximum load sustained by the workpiece during the tension test divided by its original cross sectional area. It is sometimes called the ultimate strength of the material.

（4）**Tensile Elongation**   This is frequently taken as an indicator of the ductility of the material under tension test. To determine the elongation, the increase in distance between two reference marks, scribed on the workpiece before test, is measured with the two halves of the broken workpiece held together.[②] The percentage elongation is 100 times the quotient of the increase in length and the initial distance between the scribe marks.

（5）**Reduction of Area**   This is the quotient of the decrease in cross sectional area at the plane of fracture and the original area at that plane (times 100, to express as a percentage). Similarly to percentage elongation, this number is related to ductility.

Questions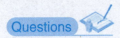

1. What can be calculated from the load-elongation or stress-strain curve?
2. What is the tensile strength?
3. How to determine the elongation?

Notes

①When a material under tension reaches the limit of its elastic strain and begins to flow plastically, it is said to have yielded.

当材料在拉力的作用下达到它的弹性应变极限时，开始产生塑性流变，这就叫作屈服。

句中 it 作形式主语，when 引导的状语从句作真正的主语。

②To determine the elongation, the increase in distance between two reference marks, scribed on the workpiece before test, is measured with the two halves of the broken workpiece held together.

为了确定伸长率，拉伸前在工件上标出参考标距，将拉断后工件的两部分拼在一起，测量其标距，标距的增加量即伸长量。

在本句中，the increase in distance between two reference marks 作 to determine the elongation 的同位语；而 scribed on the workpiece before test 则作 marks 的定语。

拉伸试验

结构和系统的设计需要确定零件的几何尺寸，同时（结构和系统的设计）要以有适当力学性能的材料为基础。

拉伸试验　拉伸试验是最常用的测试材料力学性能的方法。下图是典型纯金属的力-伸长量曲线。

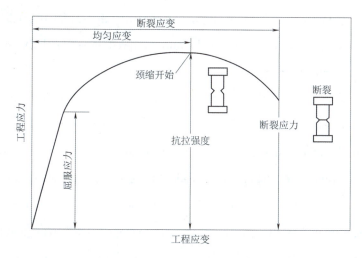

图 2-1　典型纯金属的力-伸长曲线

（1）**弹性模量**　弹性模量是指材料发生弹性变形时应力除以应变的比值，也是应力-应变曲线中直线的斜率。

（2）**屈服强度**　当材料被拉伸时弹性应变达到极限后，开始发生塑性流变，这就叫作材料屈服。屈服强度就是材料开始发生塑性流变时的应力。

（3）**抗拉强度**　抗拉强度是指工件在做拉伸试验时所能承受的最大力除以原始横截面面积的比值，有时也称为材料的极限强度。

（4）**伸长率**　伸长率通常被认为是材料在拉伸试验条件下的塑性指标。为了确定伸长率，拉伸前在工件上标出参考标距，将拉断后工件的两部分拼在一起，测量其标距，标距的增加量即伸长量。伸长量与原始标距比值的 100 倍即伸长率的百分数。

（5）**断面收缩率**　断面收缩率是工件拉断后横截面面积的缩小量与原始截面面积的商（商的 100 倍表示成百分数）。它与伸长率的百分数相似，同样是与塑性相关的指标。

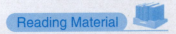

Compression Testing

Compression testing is an extremely valuable testing procedure that is often overlooked because it is not properly understood. One of the main advantages of the compression test is that tests can be performed with a minimum of material, and thus mechanical properties can be obtained from workpieces that are too small for tension testing. Compression tests are also very helpful for predicting the bulk formability of materials (behavior in forging, extrusion, rolling, etc.).

In compression testing, the material does not neck as in tension, but undergoes barreling; failure occurs by different mechanisms and therefore there is no ultimate tensile strength (UTS). In general, ductile materials do not fail in compression but tend to flow in response to the imposed loads. Brittle cylindrical workpieces loaded in compression fail in shear on a plane inclined to the load, and therefore actually break into two or more pieces. In this case, an ultimate (compressive) stress can be defined.

In comparison with tension testing, several difficulties are encountered in conducting compression tests and interpretation of the experimental data. For example, maintaining complete axiality of the applied load is important. In tension testing, self-aligning grips make this relatively simple to accomplish. In compression testing, if the workpiece is tall in relation to its diameter, this can present a major difficulty.

Hardness Test

The hardness test measures the resistance to penetration of the surface of a material by a hard object. A variety of hardness tests have been devised, but the most commonly used are the Brinell hardness and the Rockwell hardness test.

In the Brinell hardness test, a hard steel sphere, usually 10mm in diameter, is forced into the surface of the material. The diameter of the impression left on the surface is measured and the Brinell hardness number (BHN) is calculated from the following equation.

$$BHN = \frac{F}{(\pi/2)\ D\ (D-\sqrt{D^2-D_i^2}\ )}$$

where F is the applied load measured in Newton; D is the diameter of the indentor measured in millimeters, and D_i is the diameter of the indentation measured in millimeters.

The Rockwell hardness test uses either a small diameter steel ball for soft materials or a diamond cone, or brale, for harder materials. The depth of penetration of the indentor is automatically measured by the testing machine and converted to a Rockwell hardness number.

The Vickers and Knoop hardness tests are microhardness tests; they form such small indentations that a microscope is required to obtain the measurement.

The hardness numbers are used primarily as a basis for comparison of materials, specifications for manufacturing and heat treatment, quality control, and correlation with other properties and behavior of materials. For example, Brinell hardness is closely related to the tensile strength.

A Brinell hardness number can be obtained in just a few minutes with virtually no preparation of the workpiece and without destroying the component, yet provides a close approximation for the tensile strength.

Hardness correlates well with wear resistance. A material used to crush or grind ore should be very hard to assure that the material is not eroded or abraded by the hard feed materials. Similarly, gear teeth in a transmission or drive system of a vehicle should be hard so that the teeth do not wear out.

New Words

1. tension   *n.* 拉伸，拉紧

2. elongation   *n.* 伸长，伸长率

3. modulus   *n.* 模量，系数，率

4. slope   *n.* 倾斜角，倾斜面，斜度，坡（梯）度

5. sustain   *vt.* 持续，支撑

6. ultimate   *n.* 极限，终极，极点

7. ductility   *n.* 韧性，塑性

8. scribe   *n*. 划线器；*vt*. 用划线器划线

9. quotient   *n*. 商数，系数

Phrases and Expressions

1. tension testing   拉伸试验

2. mechanical property   力学（机械）性能

3. load-elongation curve   力–伸长曲线

4. stress-strain curve   应力–应变曲线

5. modulus of elastically   弹性模量

6. elastic deformation   弹性变形

7. yield strength   屈服强度

8. cross sectional area   横断面积

9. ultimate strength   极限强度

10. tensile elongation   拉伸伸长率

11. reference mark   参考标记

科技英语的翻译

翻译的问题，主要在于怎样正确地理解原文和怎样确切地用译文语言表达出来。因此，翻译过程可分为理解和表达两个阶段。

第一阶段为理解阶段。理解阶段是通过英语来掌握原作的思想内容，也是翻译的一个重要阶段。正确理解原作是翻译的基础。这就要求我们在翻译时，首先必须认真阅读原文，了解大致内容和专业范围，有时还需要查阅有关的资料，待领会原作的精神时，才可下笔开译，这样就不致于出现大错。

理解原文时，必须根据英语的语法规律和习惯去理解，把原作内容彻底弄清楚。对原作的理解应包括词汇、语法和专业内容三个方面。只有这三方面都理解透彻，才能做出正确的表达。请看例句：

A stress is therefore set up between the two surfaces which may cause the glass to break.

错误译文：因而在引起玻璃破裂的两个表面之间产生了一个应力。

正确译文：因而在这两个表面之间产生了使玻璃破裂的应力。

前者产生错误的原因在于对原句的错误理解，把 which 引导的定语从句看成修饰 surfaces 的，实际上 which 引导的定语从句是修饰 a stress 的，只是为了避免头重脚轻，保持句子平衡，才把定语从句放到了后面。另外，从专业角度看，使玻璃破裂的不可能是表面，而只能是应力。由此可以看出理解的重要性。

翻译过程的第二个阶段为表达阶段。表达阶段的任务是从汉语中选择恰如其分的表达手段，把已经理解了的原文内容重述出来。如果说，在理解阶段必须"钻进去"，把原文内容吃透，那么在表达阶段就必须"跳出来"，不受原文形式的束缚，而根据汉语的语法规律和习惯来表达。

在表达阶段最重要的是表达手段的选择，也就是"跳出来"的问题。在正确理解原作的基础上，同一个句子可能有几种不同的译法，但译文质量并不相同。请比较下面的译例：

Action is equal to reaction, but it acts in a contrary direction.

译文一：作用与反作用相等，但它向相反的方向起作用。

译文二：作用与反作用相等，但作用的方向相反。

译文三：作用与反作用力大小相等，方向相反。

应该说三种译文在表达原意上都是正确的，但可以看出，在表达的简练、通顺上，后面的译文依次要比前面的好。

虽然翻译过程可分为理解阶段和表达阶段，但两者不是截然分开的，而是互相联系的。请看例句：

In fact, it may be said that anything that is not an animal or a vegetable is a mineral.

本句并不难理解，这是一个主从复合句，主句带有一个主语从句，而主语从句又带有一个定语从句。按语法分析，本句可译为"事实上，可以说不是动物或植物的任何东西便是矿物"。但译文不够通顺，进一步分析，便可发现，"不是动物或植物"是"任何东西便是矿物"的条件。因此，可将此句译为"事实上，可以说任何东西只要不是动物或植物便是矿物。"

这一改动，译文就通顺多了，而且更符合原文意思。由此可见，透彻理解是准确表达的前提，而通过准确表达又能达到更透彻的理解。

1. 直译与意译

（1）直译。

Physics studies force, motion, heat, light, sound, electricity, magnetism, radiation, and atomic structure.

物理学研究力、运动、热、光、声、电、磁、辐射和原子结构。

The outcome of a test is not always predictate.

试验的结果并不总是可以预料的。

（2）意译。

We can get more current from cells connected in parallel.

电池并联时提供的电流更大。

The law of reflection holds good for all surfaces.

反射定律对一切表面都适用。

2. 合译与分译

（1）合译。

合译就是把原文两个或两个以上的简单句或复合句，在译文中用一个单句来表达。

There are some metals which possess the power to conduct electricity and ability to be magnetized.

某些金属具有导电和被磁化的能力。

（2）分译。

分译就是把原文的一个简单句中的一个词、词组或短语译成汉语的一个句子，这样原文的一个简单句就被译成汉语的两个或两个以上的句子。汉语习惯用短句表达，而英语使用长句较多。在科技英语翻译过程中要把原文句子中复杂的逻辑关系表达清楚，经常采用分译法。

With the same number of protons, all nuclei（nucleus 的复数）of a given element may have different numbers of neutrons.

虽然某个元素的所有原子核都含有相同数目的质子，但是它们含有的中子数可以不同。

3. 增译与省译

（1）增译。

增译就是在译文中增加英语原文省略，或原文中无其词而有其意的词语。

The amount of cell respiration depends upon the degree of activity of the organism.

细胞呼吸量的大小由生物体活动的程度来决定。

（2）省译。

严格来说，翻译时不允许对原文的内容有任何删略。由于英汉两种语言表达方式的不同，英文句子有些词语如果直译成汉语，反而会使译文难懂。为使译文通顺、准确地表达原文的思想内容，有时需要将一些词语省略不译。

4. 顺译与倒译

（1）顺译。

顺译就是按照原文相同或相似的语序进行翻译。

The nut must not protrude above the metal surface.

螺母不应高出金属表面。

Space programs demand tremendous quantities of liquid hydrogen and oxygen as rocket fuel.

航天计划需要大量的液氢和液氧作为火箭燃料。

（2）倒译。

为了使译文通顺，采用不同于原文词语顺序的方法来翻译。

Too large a current must not be used.

不得使用过大的电流。

The converse effect is the cooling of a gas when it expends.

气体在膨胀时将其冷却是一种逆效应。

Chapter 3　Hardness Testing，Impact Testing and Fatigue Testing

Hardness Testing　The hardness of a metal，defined as the resistance to penetration，gives a conveniently rapid indication of its deformation behavior. The hardness tester forces a small sphere，pyramid or cone into the surface of the metals by means of a known applied load，and the hardness number (Brinell，Rockwell or Vickers) is then obtained from the quotient of the test load and the area of the metal，since during the indentation，the material around the impression is plastically deformed to a certain percentage strain. [1]

Impact Testing　A material may have a high tensile strength and yet be unsuitable for shock loading conditions. To determine this the impact resistance is usually measured by means of the notched or un-notched izod or Charpy impact test (Fig. 3 – 1). In this test giving a load pendulum from a given height to strike the workpiece，and the energy dissipated in the fracture is measured. The energy absorption is supposed to reveal the toughness of metals under impact loading conditions.

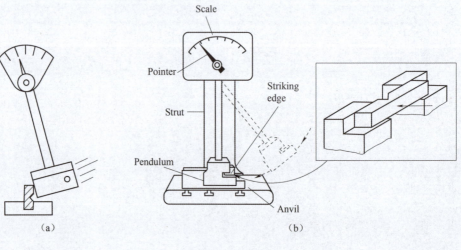

Fig. 3–1　Impact test

(a) Izod;　(b) Charpy

Fatigue Testing Fatigue testing determines the ability of a material to withstand repeated applications of a stress, which in it is too small to produce appreciable plastic deformation. Several different types of testing machines have been constructed in which the stress is applied by bending, torsion, tension or compression, but all involve the same principle of subjecting the material to constant cycles of stress. [2] To express the characteristics of the stress system, three properties are usually quoted, which include ①the maximum range of stress, ②the mean stress, and ③the time period for the stress cycle. The standard method of studying fatigue is to prepare a large number of workpieces free from flaws and to subject them to tests using a different range of stress, S, on each group of workpieces. The number of stress cycles, N, endured by each workpiece at a given stress level is recorded and plotted. This S-N diagram (Fig. 3-2) indicates that some metals can withstand indefinitely the application of a large number of stress reversals, provided the applied stress is below a limiting stress known as the endurance limit (σ_r). [2]

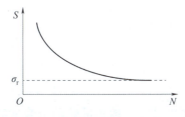

Fig. 3-2   *S-N* fatigue curve

Questions

1. Is a material which may have a high tensile strength suitable for shock loading conditions? If so, why?

2. How are the characteristics of the stress system expressed during fatigue testing?

3. What is the hardness of a material?

4. What does the S-N diagram indicate?

Notes

① ... since during the indentation, the material around the impression is plastically deformed to a certain percentage strain.

……由于在压头压入金属的过程中，压痕周围的材料会发生一定量的塑性变形。本句为原因状语从句。

②Several different types of testing machines have been constructed in which the stress is applied by bending, torsion, tension or compression, but all involve the same principle of subjecting the material to constant cycles of stress.

已经研制出了几种不同类型的试验机，用它们可以进行弯曲、扭转、拉伸、压缩等试验，但它们都有一个相同的原理，即材料承受恒定的循环应力。

句中 in which 引导定语从句作 testing machines 的定语。

③This *S-N* diagram（Fig. 3-2）indicates that some metals can withstand indefinitely the application of a large number of stress reversals, provided the applied stress is below a limiting stress known as the endurance limit.

这个 *S-N* 曲线（图3-2）表明，对于一些金属而言，如果所承受应力低于它能承受的极限应力时，它就可以承受无限次应力循环。

句中由 provided 引导让步状语从句；而 known as 分词短语作 a limiting stress 的定语。

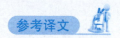 参考译文

硬度试验、冲击试验和疲劳试验

硬度试验　金属的硬度就是抵抗硬物压入的能力。它给了金属变形行为一个简便而迅速的指标。硬度试验是用一个棱锥或圆锥小球以已知的载荷施加于金属表面，由于在压头压入金属的过程中压痕周围的材料会发生一定量的塑性变形，所以硬度值（布氏、洛氏、维氏）是试验载荷除以金属受载荷面积的商。

冲击试验　一种材料可能有很高的强度，却不适合在冲击载荷条件下使用。为了确定它的冲击抗力，常通过缺口或无缺口或夏比冲击试验来测定（图3-1）。在冲击试验中，给摆锤一个高度去敲击工件，可测得工件断裂时损失的能量。损失的能量被认为是在冲击载荷条件下金属的韧性。

疲劳试验　疲劳试验是用来确定材料承受循环应力的能力，这种应力太小而不会产生大量的塑性变形。已经研制出了几种不同类型的实验机，用它们可以做弯曲、扭转、拉伸、压缩等试验，但它们都有一个相同的原理，即材料承受恒定的循环应力。为了描述此应力系统的特征，通常有三项特征描述：①应力幅，②名义应力，③应力周期。研究疲劳的标准方法是准备大量的无缺陷工

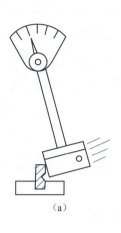

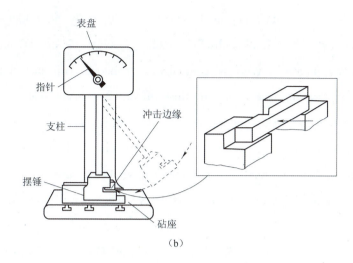

图 3-1　冲击试验

（a）缺口；（b）夏比

件，让每一组工件受不同的应力，即 S，每个工件在给定应力水平下所标记和测量的应力周期次数为 N。这个 S-N 曲线（图 3-2）表明，对于一些金属而言，如果所承受应力低于它能承受的极限应力（σ_r）时，它就可以承受无限次应力循环。

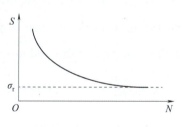

图 3-2　S-N 疲劳曲线

Reading Material

Creep Test

The flow or plastic deformation of a metal held for long periods of time at stresses below the normal short-time yield strength is known as creep. Although we normally think of creep as occurring only at elevated temperatures, room temperature can be high enough for creep to occur in some metals. In lead, for example, creep at room temperature is common. In many cases, lead pipes must be supported to prevent sagging under their own mass.

All metals creep under load at sufficiently high temperatures. At temperatures below 40% of the absolute melting point, creep is normally not a problem; consequently, creep is generally of concern with materials subjected to elevated temperatures and is a long-time effect. For plastic materials, the creep is often a problem at room temperature or at slightly elevated temperatures. For metals creep usually becomes a problem at relatively high service temperature. Many high-temperature, creep-resistant alloys have been developed for used in steam and gas turbines, high-temperature pressure vessels, power plants in general, and so on.

Brittle Behavior and Ductility of Materials

Brittle Behavior Ductile materials display an engineering stress-strain curve that goes through a maximum at the tensile strength. In more brittle materials, the maximum load or tensile strength occurs at the point of failure. In extremely brittle materials, such as ceramics, the yield strength, tensile strength, and breaking strength are all the same.

Ductility Ductility measures the amount of deformation that a material can withstand without breaking. There are two ways to express ductility. First, we could measure the distance between the gage marks on our workpiece before and after the test. The elongation describes the amount that the workpiece stretches before fracture.

$$\text{Elongation} = \frac{l_f - l_0}{l_0} \times 100\%$$

where l_f is the distance between gage marks after the workpiece breaks.

A second approach is to measure the percent change in cross sectional area at the point of fracture before and after the test. This reduction in area describes the amount of thinning that the workpiece undergoes during the test.

$$\text{Reduction in area} = \frac{A_0 - A_f}{A_0} \times 100\%$$

where A_f is the final cross sectional area at the fracture surface.

Ductility is important to both designers and manufactures. The designer of a component would prefer a material that displays at least some ductility so that, if the applied stress is too high, the component deforms before it breaks. A fabricator wants a ductile material so he can form complicated shapes without breaking the material in the process.

New Words

1. penetration *n.* 压入
2. pyramid *n.* 锥体
3. cone *n.* 圆锥
4. indentation *n.* 压痕，压坑，印压
5. impression *n.* 压痕
6. notch *n.* 缺口
7. pendulum *n.* 摆锤
8. fatigue *n. & v.* 疲劳
9. torsion *n.* 扭转；转矩

Phrases and Expressions

1. hardness testing 硬度试验
2. Brinell hardness 布氏硬度
3. Rockwell hardness 洛氏硬度
4. Vickers hardness 维氏硬度
5. impact testing 冲击试验
6. impact loading 冲击载荷
7. fatigue testing 疲劳试验
8. plastic deformation 塑性变形
9. endurance limit 疲劳极限，持久极限

科技英语常用词缀

1. 前缀

按构词的功能可分为基本前缀和术语前缀。

基本前缀主要是词缀法派生单词用的构词词缀。

术语前缀的特点就是这类词头本身是词根，可以和各个不同意义的词根结合成各种专业术语。

2. 后缀

后缀按构词的功能也可分为基本后缀和术语后缀。

术语后缀可分为一般性术语后缀和专业术语后缀。带有一般性术语后缀的词汇在科技书籍或文献中出现较为频繁。它们大部分在各个学科中是通用的。

专业性术语后缀只能构成专业的科技词汇，如生物、医学、化学等的科技词汇里就有许多专业性术语后缀。

3. 常出现的术语前缀

aniso（anis）-　（词头）不同（等，均），参差，非等同

anisotropy　各向异性，非均质性

anisoelasticity　非（等）弹性的

de-　去，消，减，分，离，防，反，等

demagnetise　退磁，消磁

deoxidize　去氧化

electro-　电（气，化，动，解）

electroconductivity　电导率，导电性

electromagnetic 电磁的

ferro-　（词头）铁的，铁合金的

ferromagnetic　铁磁性的，铁磁体

ferrocobalt　铁钴合金

homo-　相同，均匀，相似，同质

homogeneity　均匀性，一致性，同质性

homomorphic　同形的，同态的

hydro-　（词头）水，流体，氢化

hydrophily　亲水性

hydrogenate　使氢化，加氢，氢化物

infra-　（词头）下，次，亚，低，外

infrared　红外线的，红外辐射的

infrastructure　下部结构，底层结构

inter-　（词头）在中间，在内，相互

interaction　相互作用

intercrystalline　结晶内的，晶粒间的

iso-　同等，均匀，异构

isotope　同位素

isotrope　各向同性，匀质

micro-　微小，细，百万分之一

micrometer　测微器，千分尺

micrometre　微米

morpho-　形状，形态

morphology　结构，形态（学），表面几何形状

morphometry　形态测量学

mono-　（词头）单，一，单一的

monoxide　一氧化物

monocrystal　单晶体

mult（i）-　（词头）

multifrequency　多频的，宽带的

multifunction　多功能

photo-　（词头）光，光电，照相

photometer　光度计，曝光表

photosensitive　光敏的，感光的

polari-　（词头）极

polarity 极性，极化，偏光性

polarization 极化（作用），两极分化

poly- （词头）多，聚，重，复

polyporous 多孔的

polyphase 多相的

proto- 第一，初始，原始，首要

prototype 原型，样机，模型机

protogenous 原生的

pseudo- 假，伪，赝，拟，准

pseudoimage 假象

pseudostress 伪应力

radio- 放射，辐射，无线电，X 射线

radiography（RT） 射线照相术，放射照相术（RT）

radiograph 射线照片，放射照片

semi- 半，部分，不完全

semiconductor 半导体

semidiameter 半径

thermo- 热（电），温；therm- 热（电）

thermometer 温度计

thermoelasticity 热弹性

trans- 横过，贯通，超越，转换

transducer 换能器，传感器

transection 横断面

ultra- 超，越，限外，在……的那一边

ultrasonic　超声波的，超音速的

ultraviolet　紫外的，紫外线的，紫外辐射

kilo-　千

kilometer　千米

kilogram　千克

用以表示十进制倍数的词缀及符号（以 metre（米）为例，直接组合）

词缀	符号	中文	数值	词缀	符号	中文	数值
tera	T	太（拉）	10^{12}	centi	c	厘	10^{-2}
giga	G	吉（咖）	10^{9}	milli	m	毫	10^{-3}
mega	M	兆	10^{6}	micro	μ	微	10^{-6}
kilo	K	千	10^{3}	nano	n	纳	10^{-9}
hecto	h	百	10^{2}	pico	p	皮（可）	10^{-12}
deca	da	十	10	femto	f	飞（母托）	10^{-15}
deci	d	分	10^{-1}	atto	a	阿（托）	10^{-18}

4. 常出现的术语后缀

-graphy 表示"……方法，术，学"

radiography　射线照相术

thermography　红外照相术

-graph 表示"书写、画、记录所用的器械、工具，图表"

radiograph　射线照片

bolograph　辐射计

-gram 表示"图像、图表、书写物、记录"

radiogram　射线照片

tomogram　层析图

-ium 表示"金属元素及气体名称"

aluminium　铝

germanium　锗

caesium　铯

thulium　铥

ytterbium　镱

iridium　铱

selenium　硒

helium　氦

-meter 表示"计量器具，仪表，……仪，……计"

magnetometer　磁力仪

hardometer　硬度计

-metry 表示"测量学，度量学，测量方法"

diffractometry　衍射测定法

thermometry　温度测定法

-ode 表示"电极，……极管"

anode　阳极

cathode　阴极

-on 表示"原子结构的物质成分"

electron　电子

neutron　中子

-scopy 表示"观察、检测的技术、方法"

microscopy　显微术，显微学

radioscopy　射线透视

-scope 表示"观察的仪器，……镜，指示器"

microscope　显微镜

telescope　望远镜

Chapter 4　The Fe-Fe₃C Phase Diagram

The Fe-Fe$_3$C phase diagram provides the basis for understanding the ferrous alloys.

Solid Solutions　There are three solid solutions of importance δ-ferrite, γ-austenite, and α-ferrite and one intermetallic compound—cementite (Fe$_3$C). In addition, a metastable phase—martensite can form on rapid cooling.

Three-phase reactions　The three-phase reactions in Fe-C phase diagram, namely peritectic reaction, eutectic reaction and eutectoid reaction are shown below:

Peritectic: L0. 53%C+δ0. 09%C $\longrightarrow$γ0. 17%C

Eutectic: L4. 3%C $\longrightarrow$γ2. 11%C+Fe$_3$C6. 67%C

Eutectoid: γ0. 77%C $\longrightarrow$α0. 0218%C+Fe$_3$C 6. 67%C

Microconstituents　Several microconstituents may form depending on how we control the eutectoid reaction. Pearlite is a lamellar mixture of ferrite and cementite. Bainite is a nonlamellar mixture of ferrite and cementite obtained by transformation of austenite at a large undercooling. Either primary ferrite or primary cementite may form, depending on the original composition of the alloy. Tempered martensite, a mixture of very fine cementite in ferrite, forms when martensite is reheated. [1]

All of the heat treatments of a steel are directed towards producing the mixture of ferrite and cementite which gives the proper combination of properties. [2] The next few section will describe some of the techniques by which we can exercise control in unalloyed steels.

We will concentrate on the eutectoied portion of the Fe-Fe$_3$C phase diagram (Fig. 4–1).

The solubility lines and the eutectoid isotherm are identified by A_3, A_{cm}, and A_1. The A_3 shows the temperature at which ferrite starts to form on cooling; the A_{cm} shows the temperature at which cementite starts to form; and the A_1 is the eutectoid temperature.

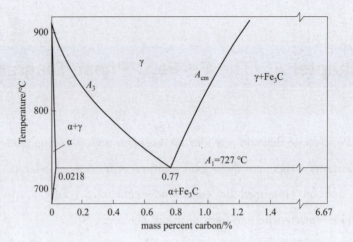

Fig. 4-1 The eutectoid portion of the Fe-Fe₃C phase diagram

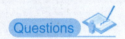

1. What is an equilibrium phase diagram?

2. What are liquidus line and solidus line?

3. How does the solubility of one metal in another generally change with a decrease in temperature?

4. What are the three-phase reactions that appear in the iron-carbon phase diagram?

①Tempered martensite, a mixture of very fine cementite in ferrite, forms when martensite is reheated.

回火马氏体，是细小渗碳体分布在铁素体上的混合物，只有当马氏体被重新加热后才会形成。

句中 a mixture of very fine cementite in ferrite 为同位语，进一步说明 tempered martensite。

②All of the heat treatments of a steel are directed towards producing the mixture of ferrite and cementite which gives the proper combination of properties.

钢的所有热处理均倾向于形成铁素体和渗碳体组成的具有特殊性能的混

合物。

句中 all 作名词用，all of 可译为"所有的；全部"等；which gives the proper combination of properties 为定语从句，说明 the mixture of ferrite and cementite，可译为"具有特殊性能"。

参考译文

铁–碳合金相图

铁–碳相图为人们研究铁–碳合金奠定了基础。

固溶体　铁具有三种同素异晶结构，即 δ-Fe，γ-Fe 和 α-Fe，还有一种金属间化合物，即渗碳体（Fe_3C）。另外，在快速冷却条件下会形成一种亚稳定相，即马氏体。

三种相反应　铁–碳相图中有三种相反应，即包晶反应、共晶反应和共析反应，其反应式如下。

包晶反应：$L0.53\%C+\delta0.09\%C \longrightarrow \gamma0.17\%C$

共晶反应：$L4.3\%C \longrightarrow \gamma2.11\%C+Fe_3C6.67\%C$

共析反应：$\gamma0.77\%C \longrightarrow \alpha0.0218\%C+Fe_3C\ 6.67\%C$

钢和铸铁在含碳量上的分界点为 2.11%，即钢的含碳量不超过 2.11%，含碳量大于 2.11% 的为铸铁。另外，此处发生共晶转变也是一个标志。

微观组织　经共析转变会形成几种微观组织。珠光体为铁素体和渗碳体的片层状结构。在大的过冷度下由奥氏体转变得到的渗碳体和铁素体的非片层状混合物即贝氏体。冷却过程中是形成铁素体还是形成渗碳体，这取决于合金最初的化学成分。回火马氏体，是细小渗碳体分布在铁素体上的混合物，只有当马氏体被重新加热后才会形成。

钢的所有热处理均倾向于形成铁素体和渗碳体组成的具有特殊性能的混合物。下面几部分描述了可用于非合金钢热处理中的一些技巧。

图 4-1 为铁–碳相图中共析转变的一部分。

图中溶解线和共析等温线可由 A_3，A_{cm} 和 A_1 来区分。A_3 线代表在冷却过程中由奥氏体析出铁素体的开始线；A_{cm} 线代表在此温度析出二次渗碳体，A_1 线代表共析反应温度。

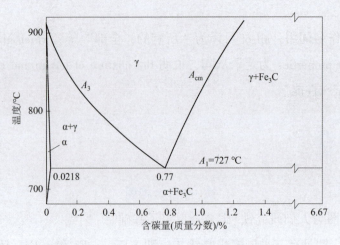

图 4-1　铁-碳相图共析部分

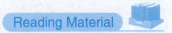

Equilibrium Diagram of Iron and Iron Carbide

Carbon is the basic for the wide range of properties obtainable in iron and steel. It forms different compositions with iron when combined in different ways and amounts. Thus carbon is the primary means for making iron or steel soft and ductile, tough or hard. Most other alloying elements in effect modify or enhance the benefits of carbon.

The fundamental effects of carbon on iron are shown in iron and carbon phase diagram of Fig. 4-2 and iron and carbide equilibrium diagram of Fig. 4-3. The carbon commonly appears as iron carbide but not always, e. g., graphite in cast iron. Iron carbide consists of 6. 67% carbon and 93. 33% iron by mass and has the average formula Fe_3C. It is the hardest constituent in carbon iron and carbon steel and is quite brittle and white in color. It is called cementite.

Iron and iron carbide solid solutions. The line for zero carbon in Fig. 4-2 shows that pure iron or ferrite solidifies at about 1,540 ℃. Over a short range of high temperature it has a body-centered-cubic structure called delta iron. When cooled to 1,400 ℃, the structure changes to face-centered-cubic gamma iron. This is the main part of austenite, which may dissolve up to 1. 7% carbon as iron carbide. Below 910 ℃, ferrite transforms to body-centered-cubic alpha iron. Thus iron is allotropic. It also changes from a nonmagnetic material at the high temperature to a magnetic one somewhat below 800 ℃. Alpha iron dissolves only a small amount of carbide. Ferrite can also hold such elements as nickel, silicon, phosphorus, and sulfur in solution in amounts depending on temperature.

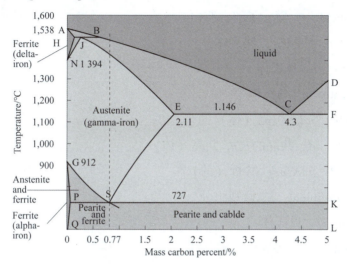

Fig. 4-2　Iron and carbon phase diagram

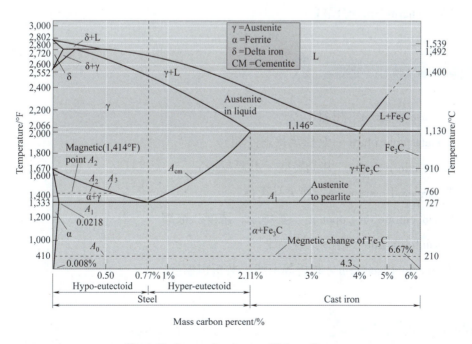

Fig. 4-3　Iron and carbon equilibrium diagram

New Words

1. provide   *v.* 供应，供给，准备，预防，规定

2. understand   *v.* 懂，了解，明白

3. ferrite   *n.* 铁酸盐，铁素体

4. austenite   *n.* 奥氏体

5. martensite   *n.* 马氏体

6. peritectic   *adj.* 包晶（体）的

7. eutectic   *adj.* 共晶的；*n.* 共晶（体）

8. eutectoid   *n. & adj.* 共析（的）

9. microconstituent   *n.* 微观组分

10. bainite   *n.* 贝氏体

11. cementite   *n.* 渗碳体，碳化铁

12. undercool   *v.* （＝supercool）（使）过度冷却，过冷

13. concentrate   *v.* 集中，浓缩

14. isotherm   *n.* 等温线

Phrases and Expressions

1. phase diagram   相图

2. metastable phase   亚稳定状态（相）

3. eutectic reaction   共晶转变（共晶反应）

4. tempered martensite   回火马氏体

5. concentrate on（upon）   集中于，浓度，浓缩

6. solubility line   溶解，溶解线

7. eutectoid isotherm   共析等温线，共析恒温线

Glossary of Terms

1. ferrous alloy   铁质合金

2. solid solution   固溶体

3. intermetallic compound　金属间化合物

4. metastable phase　亚稳定相

5. eutectic reaction　共晶反应

6. eutectoid reaction　共析反应

7. cementite　渗碳体

8. iron-carbon phase diagram　铁–碳相图

9. equilibrium phase diagram　平衡相图

科技英语中的as结构的特点和翻译

as 结构

1. as...as+数字

这一结构要译成"……达……"或"……至……"。

The temperature at the sun's center is as high as 10,000,000 ℃.

太阳中心的温度高达 10 000 000 ℃。

2. as...as+形容词（副词)

"as A as B"这一结构要译成"既 B 又 A"或"又 B 又 A"。

Alloys are as important as useful.

合金既有用又重要。

3. as much（many)...as

这一结构要译成"……（那样多）的……"或"……的……都"。

A saturated solution contains as much solid as it can dissolve.

饱和溶液含有它所能溶解的最大量的固体。

4. the same as

这一结构要译成"和……的……一样"或"与……的……相同"。

The tested conditions are the same as will be encountered in use.

测验条件和使用时要遇到的条件一样。

5. such as

这一结构要译成"……的一种……"。

The metric system is such as logical link between its units.

公制是各单位之间具有逻辑关系的一种度量衡制。

6. as..., so...

这一结构要译成"正如……一样，……也……"。

As water is the most important of liquids, so air is the most important of gases.

正如水是最重要的液体一样，空气也是最重要的气体。

7. as (so) far as... is concerned

这一结构要译成"就……而论"。

As far as construction is concerned, the computer is similar to the human brain.

就结构而论，计算机和人脑有相似之处。

8. 形容词（分词、副词）+as+主语+动词

这是一种倒装结构，as 在这里相当于 though，表示让步意义。它的语气比 though 更为强烈，表现更有力。一般译成"虽然（尽管）……，但是……"。

Complicated as the problem is, the electronic brain can solve it in a short period of time.

这个问题虽然复杂，但电脑能在很短的时间内把它解决。

Chapter 5   Metallography

The preparation of metallographic workpieces described in this article is also applicable to other types of studies, such as electron microscopy, microhardness testing, image analysis, and electron microprobe analysis. Preparation of metallographic workpieces generally requires five major operations：①sectioning, ②mounting (optional), ③grinding, ④polishing and ⑤etching (optional).

Sectioning   Many metallographic studies require more than one workpiece. For example, a study of deformation in wrought metals usually requires two sections—one perpendicular to, and the other parallel to, the direction of deformation. Bulk samples for sectioning may be removed from larger pieces or parts using methods such as core drilling, band or hack sawing, flame cutting, etc. However, when these techniques are used, precautions must be taken to avoid alteration of the microstructure in the area of interest. In general, the size of the workpiece is $\phi12\times12$ if it is cylinder or $12\times12\times12$ if it is cubic.

Mounting   Once cut, the sample piece is either ground by hand, or if the sample is small, mounted before grinding. The method of mounting should in no way be injurious to the microstructure of the workpiece. Compression mounting, the most common mounting method, uses pressure and heat to encapsulate the workpiece with a thermosetting or thermoplastic mounting material. [1]

Grinding   The purpose of grinding is to lessen the depth of deformed metal to the point where the last vestiges of damage can be removed by a series of polishing steps. A satisfactory fine grinding sequence might involve grit sizes of 240, 320, 400 and 600 grit. Movement of the workpiece at right angles to the previous grind (or polishing) will make it easier to detect the removal of marks left from previous processing. [2]

Polishing   Polishing is the final step in producing a deformation-free surface that is flat, scratch-free, and mirrorlike in appearance. After rough grinding, preliminary polishing with successively finer and finer emery paper is the next step. The process can be finished either by hand or by mechanism. Fine polishing is accomplished with wet rotating discs. These discs are covered with polishing cloths that are charged with an abrasive powder suspended in water.

Etching. After polishing samples should show only impurities. Etching makes the microstructure visible to the microscope. There is a lot of etching reagents. In general, the plain carbon steels employ 4% HNO_3 in ethyl alcohol.

Once the samples have been prepared, they may be viewed through a microscope. It has a capability of magnifying structures by as much as 2,000 times. High magnifications of this order generally require the use of oil-immersion lenses.

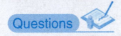 Questions

1. Describe the procedure for preparing a metallographic workpiece.

2. What size of the workpiece sectioned?

3. Why should the sample be mounted?

4. When grinding, what should you pay attention to?

 Notes

①Compression mounting, the most common mounting method, uses pressure and heat to encapsulate the workpiece with a thermosetting or thermoplastic mounting material.

压力镶嵌是一种最常用的镶嵌方法，它通过加压和加热的方式把工件用热固性和热塑性材料封装起来。

句中 the most common mounting method 作 compression mounting 的同位语。

②Movement of the workpiece at right angles to the previous grind (or polishing) will make it easier to detect the removal of marks left from previous processing.

工件研磨是垂直于前一道工序方向进行的，这样可以更容易地检验前一道研磨（或抛光）工序残留划痕去除的情况。

句中 it 作形式宾语，动词不定式 to detect the removal of marks left from previous processing 作真正的宾语。

参考译文

金相学

本文所叙述的金相工件的制备方法也适用于其他方面的研究，比如，电子显微学、显微硬度测试、影像分析和电子探针分析等。金相工件的制备通常需要五个主要步骤：①取样，②镶嵌（可选择的），③研磨，④抛光和⑤腐蚀（可选择的）。

取样 许多金相研究需要的工件不止一个。例如，金属在锻造时的变形研究，通常需要在两个截面上取样——一个是与变形方向垂直的方向，另一个则是与变形方向平行的方向。较大的部件或零件可以用孔钻、带锯或钢锯及火焰切割等方法取成块状工件，无论用哪种方法取样，应尽量避免取样部位显微组织的变化。通常，圆柱工件的取样尺寸为 $\phi12×12$，立方体工件的取样尺寸为 $12×12×12$。

镶嵌 工件一旦切割后，在研磨前若手握光滑或比较小时应镶嵌。镶嵌的方法绝不能损坏工件的显微组织。压力镶嵌是一种最常用的镶嵌方法，它通过加压和加热的方式把工件用热固性或热塑性材料封装起来。

研磨 研磨的目的是采用一系列的抛光步骤把金属变形层的深度减小到能够消除的程度。理想的研磨顺序是砂纸的规格依次为 240、320、400 和 600 号。工件研磨是垂直于前一道工序方向进行的，这样可以更容易地检验前一道研磨（或抛光）工序残留划痕去除的情况。

抛光 抛光是制备一个平整的、无划痕的、镜面似的、无变形层表面的最后一道工序。在粗磨后，下一道工序是辅助抛光（细磨），即依次用更细的金刚石砂纸去磨光，这道工序可用手工磨光也可用机械磨光来完成。细抛是在湿的旋转盘上完成的，这些旋转盘上覆有抛光布，同时在抛光布上喷洒研磨剂粉末悬浮液。

腐蚀 抛光完后的样件仅可显示夹杂物。腐蚀可使其显微组织在显微镜下看得见。腐蚀剂种类很多，最常用的是 4% 的硝酸酒精溶液。

工件制备好后，可在显微镜下观察。显微镜的放大倍数可达 2 000 倍左右，高放大倍数的显微镜通常需要用油浸透镜。

Reading Material

Microscopic Techniques and Science

Microscopic techniques and science of metallography have many advances. Some decades years ago, a magnification of 1,500×was reckoned as high, then it become possible to photograph the structures of metals at magnifications of up to 7,000×by use of ultraviolet light. Now with the electron microscopes of nearly 100,000×are possible. The electron microscope is an instrument that uses a high-velocity electron beam as an energy source.

Whereas optical microscopes use lenses as light gathering and electron beam, the electron microscopy uses a magnetic field to control and focus an electron beam. Because of the very much shorter wavelengths of electron waves as contrasted with light waves, the resolution of an electron microscope is very much higher than that of the optical microscope. Magnifications are of the order of 35,000 diameters, and may be increased when accessories are used to 200,000 diameters.

The electron microscopes are divided into two general types—transmission electron microscope and scanning electron microscope. The scanning electron microscope has become widely used for studying the fractured surfaces of metals and alloys, and from that information the resistance of metals to stresses has been improved. The scope of this particular type of microscope has also been increased further by fitting a head which analyze the energy dispersed from the electron beam, giving a rapid way of locating the elements present.

New Words

1. workpiece   *n.* 工件
2. section   *vt.* 截，剖；*n.* 截面，切断
3. mount   *v.* 镶嵌，固定
4. grind   *v.* 研磨
5. etch   *v. & n.* 腐蚀（剂），侵蚀（剂）
6. encapsulate   *v.* 密封（闭），封装，灌封
7. thermosetting   *adj.* 热固（凝）的

8. thermoplastic　*adj.* 热塑（性）的

9. vestige　*n.* 形（痕，遗）迹，残余（留）

10. grit　*n.* 粒度

11. emery　*n.* 金刚砂，刚石粉

12. disc　*n.* 圆盘

13. abrasive　*adj.* 磨蚀的；*n.* 磨蚀剂，研磨料

14. magnify　*v.* 放大，扩大

Phrases and Expressions

1. metallographic workpiece　金相工件

2. microhardness testing　显微硬度实验

3. electron microprobe analysis　电子显微探针分析

4. core drilling　孔钻

5. band sawing　带锯

6. hack sawing　钢锯

7. a series of　一系列

8. scratch-free　无划痕的

9. etching reagent　腐蚀试剂

10. ethyl alcohol　乙醇

11. oil-immersion lense　油浸透镜

12. ultraviolet light　紫外线光

13. electron microscopy　电子显微镜检查

14. optical microscope　光学显微镜

15. transmission electron microscopy　透射电子显微镜

16. scanning electron microscope　扫描电子显微镜

17. electron beam　电子束

科技英语中的it结构的特点和翻译

it 结构

1. it follows that

在这一结构中，that 从句是主语从句，it 是引导词作形式主语，follow 是不及物动词表示"归结"的意思。可译成"由此可见""由此得出"等。

It follows that the greater conductance a substance has, the less the resistance is.

由此得出，一种物质的电导率越大，电阻越小。

2. it is+形容词+for……+不定式

在这一结构中，it 是形式主语，for 以及其后的不定式是不定式复合结构，for 引出的是不定式的逻辑主语。

It is necessary for us to know how to convert energy.

我们必须弄清楚能量是怎样转换的。

3. it is no use+动名词

在这一结构中，it 是形式主语，实际主语是后面的动名词短语。

It is no use learning without practice.

学习理论而不实践是无用的。

4. it is（was）+not until...that

在这一结构中，it 不是形式主语，而是用来加强语气。习惯上译成"直到……才……"。

It was not until 1886 that aluminum came into wide use.

直到 1886 年铝才得到广泛的应用。

5. it is...since

这也是一种强调时间状语的结构，但与上一种结构不同。这里 it 不是引导词而是无人称代词，表示时间作主句的主语，连词 since 引出时间状语从句。一般译成"……已经……"或"自从……以来"。

It is ten years since the old scientist has been working at this problem.

这位老科学家研究这一题目已经整整 10 年了。

Chapter 6　Heat Treatment

Because it covers a variety of complex processes, heat treatment is not easily described. There are many metalworking procedures for metal shaping or external treatment, such as rolling, forming, machining, and plating. But only heat treatment can significantly change the ultimate condition of these shapes.

Grinding can sharpen the edge on a knife, but heat treatment makes the edge stay hard and sharp.

Forging and machining can produce a crankshaft, but gives it strength and makes the journals resistant to wear.

Stamping can produce a coiled shape, but gives the part high elasticity.

What is it?

In general, heat treatment is the controlled use of time and temperature to produce predictable changes in the internal structure of metals. [1] Materials other than metals—plastics, glass and ceramics—also may be heat treated.

Without the word internal, the definition could be construed literally to include welding, flame cutting and even melting, but heat treatment in the classic metallurgical sense has to do with the atomic and the improvements that can be made by altering them. Heat treatment involves little outward physical change and keeping a solid state without melting in the whole process.

It does, however, deal with metals as solid solutions; that is, while the object being heat treated always retains its essential shape, it can be in a state of internal flux in which vast atomic rearrangements take place.

Heat treatment is performed on metals after they have been given some sort of shape, such as plate, sheet, bar or wire. Heat treatment can be the first process following the initial shaping of metal, or it can be the last process prior to direct use of the part or its incorporation into a component, or both.

Heat treatment is a broad term encompassing a wide variety of treatments intended to accomplish different ends—softening or hardening.

Pure and alloyed metals—including iron, aluminum, copper, nickel, lead, chromium, tin, titanium, steel, brass, and bronze—can be heat treated.

It's not true, however, that all metals and alloys that can be heat softened also can be heat hardened. Heat treatment is commonly misconceived as only a hardening process and is often erroneously thought applicable only to heat hardened metals. Metals can be alloyed to become heat treatable for hardening, but it's not true that because a metal is an alloy it can be hardened by heat treatment.

Although it can take many forms, annealing generally serves to soften metals. It is often used after cold working metals that is subjecting them to physical changes such as hammering, cutting and bending. This cold working can mechanically alter their crystalline structures, causing them to harden. A simple example of this is the repeated bending of a paper clip or nail until it hardens and fractures.

By the proper application of heat, as in annealing, these work-hardened metals can be made to recrystallize, making them softer, more ductile and more amenable to further manufacturing processes. [2]

Steel is not only the most commonly used metal but also the metal most subject to the kinds of end usage that necessitate heat treatment.

Questions

1. What is heat treatment?
2. What types of materials may be heat treated? Why?

Notes

①In general, heat treatment is the controlled use of time and temperature to produce predictable changes in the internal structure of metals.

一般来说，热处理是通过控制对金属的加热和冷却的时间与温度，从而使金属内部晶体结构产生预期变化的一种方法。

②By the proper application of heat, as in annealing, these work-hardened metals can be made to recrystallize, making them softer, more ductile and more amenable to further manufacturing processes.

采用适当的热处理（比如退火），可以使加工硬化后的金属产生再结晶，从而使金属变得较软、韧性更好、更适合进行切削加工。

句中"making them softer,..."动名词短语作定语，修饰 metals。

参考译文

热处理

由于热处理是一个复杂的工艺，它的概念不太好描述。有很多金属的成形和表面处理工艺，比如轧制、成形、机械加工及电镀。但是只有热处理能够最大程度地改变金属材料的性能。

磨削可以使刀刃变尖，但是热处理可以使刀刃保持高的硬度和强度。

锻造和机械加工可以加工一个曲轴，但是热处理可以使曲轴强度提高，使轴颈更加耐磨。

冲压成形可以生产卷形弹簧，而热处理可使零件具有高的弹性。

什么是热处理呢？

一般来说，热处理是通过控制金属的加热和冷却的时间与温度，从而使金属内部晶体结构产生预期变化的一种方法。除了金属，其他材料比如塑料、玻璃和陶瓷也可以进行热处理。

因为热处理与金属的原子结构有关，而且可以通过热处理改变它的原子排列情况，所以对于热处理的概念就不像焊接、火焰切割和熔炼那么容易描述。热处理不会引起材料的物理性能变化，而且在整个过程中金属都保持固态，不会熔化。

然而，对于固溶体金属来说，虽然物体在加热时一直保持它原来的形状，但是可能会处于一种不稳定的流变状态，原子排列紊乱。

金属被加工成板状、薄片状或线材后进行热处理。热处理也可以是金属材料开始成形的第一道工序，热处理也可以作为最终工序而优于零件的直接使用，同时热处理还可作为中间工序。

热处理是一个具有广泛用途的处理工序，可以达到不同的目的——可以软化金属也可以硬化金属。

纯金属和合金——包括铁、铝、铜、镍、铅、铬、锡、钛、钢、黄铜和青铜都可以进行热处理。

然而，不是所有的金属及其合金都可以通过热处理进行软化和硬化处理。

热处理通常被误认为只是一个淬火硬化工艺，经常被误认为仅仅对金属进行硬化处理。金属被合金化后可以进行热处理硬化，但并不是只有合金才可以进行热处理硬化。

虽然热处理有多种类型，但是退火工艺主要用于软化金属。退火经常在锻打、切割和弯曲等使金属材料物理性能发生变化的冷加工之后进行。冷加工改变材料的晶体结构随之使其性能发生变化，导致材料变硬。举一个简单的例子，我们对铁丝、钉子等进行反复弯曲就会使它们变硬变脆直至断裂。

采用适当的热处理（比如退火），可以使加工硬化后的金属再结晶，从而使金属变得更软、韧性更好、更适合进行切削加工。

钢不但是最常用的金属，而且也是最适合进行各种热处理后广泛使用的金属。

Heat Treatments to Increase Strength

Six major mechanisms are available to increase the strength of metals：①solid solution hardening，②strain hardening，③grain size refinement，④precipitation hardening，⑤dispersion hardening，and ⑥phase transformations. All can be induced or altered by heat treatment but not all can be applied to any given metal.

In solid solution hardening, a base metal dissolves other atoms in solid solution, either as substitutional solutions, where the new atoms occupy sites on the regular crystal lattice, or as interstitial solutions, where the new atoms squeeze into "holes" in the base lattice. The amount of strengthening depends on the amount of dissolved solute and the size difference of the atoms involved. Distortion of the host structure makes dislocation motion more difficult.

Strain hardening produces increased strength by plastic deformation under cold-working conditions.

Because grain boundaries acts as barriers to dislocation motion, a metal with small grains will tend to be stronger than the same metal with larger grains. Thus grain-size refinement can be used to increase strength, except at elevated temperatures when failure is caused by a grain-boundary diffusion-controlled creep mechanism. Grain-size refinement is one of the few processes capable of improving both strength and ductility.

Precipitation hardening or age hardening is a method whereby strength is obtained from a nonequilibrium structure produced by a three steps (solution treat-quench-and age) heat treatment.

Strength obtained from distinct second-phase particles in a base matrix is called dispersion hardening. To be effective, these second phase should be stronger than the matrix, adding strength through both their reinforcing action and by the additional barriers presented to dislocation motion.

Phase transformation strengthening involves alloys which can be heated to form a single high-temperature phase and subsequently transformed to one or more low-temperature phase upon cooling. Where phase transformation is used to increase strength, the cooling is usually rapid and the phases produced are nonequilibrium in nature.

New Words

1. significantly   *adv.* 可观地；显著地；重要地

2. grinding   *n.* 磨光

3. stamp   *vt.* 压，捣，锤击，冲压成形

4. predictable   *adj.* 可预（示，测，报）言的

5. incorporation   *n.* 结（联，掺，混）合，加（掺）入

6. misconceive   *v.* 误解，概念不清，错觉

7. erroneously   *adv.* 不真实地；失实地

8. hammer   *n.* 锤，榔头；*v.* 锤打（击，炼），锻（造）

9. crystalline   *adj.* 结晶的；*n.* 结晶体，晶体，晶粒

10. fracture   *v.* （使）破裂，（使）断裂；*n.* 断口（面），裂缝（痕）

11. ductile   *adj.* 可延伸的，可塑的，韧性的

12. amenable   *adj.* 应负责的；顺（服）从的，适合于……的（to sth）

13. necessitate   *vt.* 需要，使成为必要，以……为条件

14. crystallize   *v.* 使结晶，晶化

15. recrystallize   *v.* 使再结晶

Phrases and Expressions

1. heat treatment　热处理
2. ductile fracture　塑形破坏
3. work-hardened　加工硬化

科技英语中的than结构的特点和翻译

than 结构

than 结构在英语中除了用来连接从句外，还可与其他词组成固定结构，常见的有以下一些。

1. more... than

这一结构除了比较级的用法外，另一种用法是对两种不同的性质加以比较，而有所取舍。可把 "more A than B" 顺译成 "是 A 而不是 B" 或译成 "与其说是 B，不如说是 A"。

The division between the pure scientists and the applied scientists is more apparent than real.

理论科学家和应用科学家的区分是表面的而不是实际的。

2. more than+数字

这一结构一般译成 "……以上" "……多"。

More than 100 chemical elements are known to man; of these, about 80 are metals.

人类已知的化学元素有 100 种以上，其中有 80 种是金属。

3. less than+数字

这一结构一般译成 "…… 以下" "不到…… "。

The carbon content in mild steel is less than 0.3%.

中碳钢的含碳量在 0.3%以下。

4. less... than

在这一结构中，less 与 than 分开使用，用法与 less than 不同，它相当于

"not so... as" 结构, 译成 "比……小" "不及……" "不如……" 等。

The electrons meet less resistance when the conductor is cold than when it is hot.

电子在导体冷却时遇到的阻力比导体热时的小。

5. other than

这一结构一般译成 "除……之外" "不仅……而且还……"。

There are practical sources of heat energy other than the combustion of fossil fuels.

除了燃烧矿物燃料外, 还有一些实用的热源。

6. rather than

这一结构一般译成 "而不是"。

It is for the travelers rather than the enthusiasts that the railways must be run.

经营铁路是为了旅客而不是为了铁路爱好者。

7. would rather... than

在这一结构中, 可把 "would rather A than B" 顺译成 "宁可 A 而不愿 B", 或逆译成 "与其 B 倒不如 A"。

We would rather go on with the experiment than give it up.

我们宁愿把这项实验进行下去, 而不愿放弃不做。

8. nothing more than

这一结构一般译成 "只不过" 或 "仅仅是"。

The thermostat is nothing more than an electric switch that opens and closes itself at the proper temperature.

恒温器只不过是到达适当温度就自动开、关的电闸。

9. no more than

这一结构与上一结构相似, 一般译成 "只不过" 或 "仅仅是"。

This first orbit nearest to the nucleus contains no more than 2 electrons.

最靠近原子核的轨道仅仅含有两个电子。

10. no sooner... than

这一结构一般译成 "一……就" "刚……就" 等。

The push button had no sooner been depressed than the motor began to run.

按钮刚一压下, 发动机就开动了。

Chapter 7　Nondestructive Testing

Nondestructive Testing　Nondestructive testing（NDT）includes visual examination, liquid penetrant, magnetic particle, eddy current, ultrasonic, radiography, acoustic emission, thermal and optical methods. Space permits only a brief discussion of these most often used NDT methods. For a more complete presentation see the references listed at the end of this book.

Visual Examination　An experienced welder or inspector can detect most of the weld defects by careful examination. The following defects can be observed, undercut, overlap, surface checks, cracks, slag inclusions, penetration, and the extent of reinforcement. Some of these defects are shown in the figure bebw.

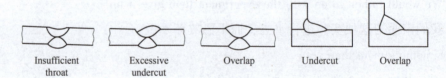

| Insufficient throat | Excessive undercut | Overlap | Undercut | Overlap |

Some welding defects that can be checked visually

Liquid Penetrant　The liquid-penetrant method involves flooding the surface with a light oil-like penetrant solution that is drawn into the surface discontinuities by capillary action. After the excess liquid has been removed from the surface, a thin coating of absorbent material is applied to draw the traces of penetrant from the defects to the surface for observation. Brightly colored dyes of fluorescent materials are added to the penetrant solutions to make the traces more visible. [1]

Simple inspection systems consist of spray can kits, including cleaner, penetrant, and developer in pressurized cans. Mass-inspection systems have been developed using fluorescent-dye penetrant, video-scanning techniques, and computer control and analysis. [2]

Questions

1. What is NDT?

2. What is visual inspection? Describe its uses.

3. Describe liquid-penetrant method.

Notes

①After the excess liquid has been removed from the surface，a thin coating of absorbent material is applied to draw the traces of penetrant from the defects to the surface for observation.

去除表面多余的液体后，为了便于观察，在表面上加一层显像剂把渗入缺陷中的渗透液吸出来。

本句中 after the excess liquid has been removed from the surface 为时间状语从句。

②Mass-production systems have been developed using fluorescent-dye penetrant，video-scanning techniques，and computer control and analysis.

利用荧光渗透剂、视频扫描技术和计算机控制与分析技术，使批量生产体系得到了发展。

由 using 引出的动名词短语，在被动语态句子中作宾语。

参考译文

无损检测

无损检测　无损检测（NDT）包括肉眼观察、液体渗透、磁粉、涡流、超声波、射线照相、声发射、热学和光学等方法。在这里仅仅对最常用的无损检测方法作一简单讨论。更多更详细的探讨参阅书后文献资料。

肉眼观察　一位有经验的焊接者或检验员能通过仔细观察检查出大多数焊接缺陷，如咬边、焊瘤、表面刻痕、裂纹、夹渣、熔深和焊缝加厚的程度，这些缺陷中的一部分如下图所示。

液体渗透　这种方法涉及一种亮得像油一样的渗透溶液流入表面，这种渗

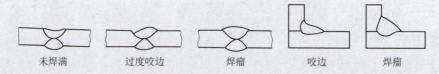

未焊满　　　过度咬边　　　焊瘤　　　咬边　　　焊瘤

可肉眼观察的焊接缺陷

透溶液是通过毛细作用进入不连续的表面中的。除掉表面多余的液体后，为了便于观察，在表面加一层显像剂把渗入缺陷中的渗透液吸出来。为了使痕迹更清晰可见，加入色彩亮丽的荧光渗透溶液。

单件检验系统是由喷洒罐等器材组成的，包括清洗剂、渗透剂和显像剂等都被压缩在喷洒罐中。批量检验系统利用荧光染色渗透剂、视频扫描技术和计算机控制与分析技术，并得到了发展。

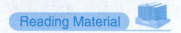
Reading Material

Definition of Nondestructive Testing

Materials, products and equipment which fail to achieve their design requirements or projected life due to undetected defects may require expensive repair or early replacement.

Such defects may also be the cause of unsafe conditions or catastrophic failure, as well as loss of revenue due to unplanned plant shutdown.

Flaws can affect the performance of materials or structures, so the detection of defects in materials, products and equipment is an essential part of quality control of engineering systems for their safe and successful use in practical situations.

Nondestructive testing is a descriptive term used for the examination of materials and components in such a way that allows materials to be examined without changing or destroying their usefulness.

Nondestructive testing is used to investigate the material integrity of the test object. The table below shows comparison of some nondestructive methods.

Comparison of some nondestructive methods

Method	Characteristics detected	Advantages	Limitations	Example of use
Ultrasonics	Changes in acoustic impedance caused by cracks, nonbonds inclusions, or interfaces	Can penetrate thick materials; excellent for crack detection; can be automated	Normally requires coupling to material either by contact to surface or immersion in a fluid such as water; surface needs to be smooth	Adhesive testing of adhesive parts, thin plate stacking, hydrogen embrittlement cracking
Radiography	Changes in density from voids, inclusions, material variations; placement of internal part	Can be used to inspect wide range of materials and thicknesses; versatile; film provides record of inspection	Radiation safety requires precautions; expensive; detection of cracks can be difficult unless perpendicular to X-ray film	Pipeline welds for penetration, inclusion, voids; internal defects in castings
Visual-optical	Surface characteristics such as finish, scratches, cracks, or color; strain in transparent materials; corrosion	very convenient; can be automated	Can be applied only to surface, through surface openings, or to transparent material	Paper, wood, or metal for surface finish and uniformity
Eddy current	Changes in electrical conductivity caused by material variations, cracks voids, or inclusions	Readily automated; moderate cost	Limited to electrically conducting materials; limited penetration depth	Heat exchanger tubes for wall thinning and cracks
Liquid penetrant	Surface openings due to crack, porosity, seams, or folds	Inexpensive, easy to use, readily portable, sensitive to small surface flaws	Flaw must be open to surface; not useful on porous materials or rough surfaces	Turbine blades for surface cracks or porosity; grinding cracks

续表

Method	Characteristics detected	Advantages	Limitations	Example of use
Magnetic particles	Leakage magnetic flux caused by surface or near-surface cracks, voids, inclusions, material or geometry changes	Inexpensive or moderate cost, sensitive both to surface and near-surface flaws	Limited to ferromagnetic material; surface preparation and post-inspection; demagnetization may be required	Railroad wheel cracks; large castings

Purposes of Nondestructive Testing

Ensuring product reliability is necessary because of the general increase in performance expectancy of the public.

NDT plays a very important part in the design of lighter, stronger, less costly and more reliable parts.

Almost every nondestructive testing method is applied in one way or another to assist in process control and so ensure a direct profit for the manufacturer.

NDT can be applied to each stage of a part's construction. Raw materials can be examined using NDT and either accepted, or rejected. Defects occur during processes like forging, welding, heat treatment, grinding can also be inspected. The in-service defects, for example fatigue cracks, corrosion, happened when the part or component are in service may be examined by NDT.

无损检测的定义

由于没有检测到缺陷会使材料、产品和设备无法达到其设计要求或预期寿命，还可能导致昂贵的维修费用或需要过早地更换零部件。

这些缺陷也可能是不安全的因素或者是造成灾难性事故的主要原因，还有可能使工厂意外停工造成额外的损失。

缺陷会影响材料或结构的性能，因此检测材料、产品和设备中的缺陷是工程质量系统控制的重要组成部分，以确保其在实际生产情况下的安全和使用。

无损检测是一个描述性术语，用于检查材料和零部件，能够在不改变或不破坏其使用性的情况下进行检查。

无损检测还可用于检查被测试对象的完整性。下表所示为常用无损检测方法的比较。

学习笔记

常用无损检测方法的比较

方法	检测原理	优势	局限性	应用举例
超声	由裂纹、未接合、夹杂物或者界面引起的声阻抗变化	能渗入厚材料；非常适合裂纹检测；能自动化	通常需要将耦合剂涂在材料表面，通过接触表面或浸入液体，如水中；要求材料表面光滑	胶粘件的黏合性检测、薄板叠层，以及氢脆裂纹
射线照相	夹杂物，气孔空隙，材料变化引起的密度变化；内部零件的放置	能够检测的材料和厚度的范围广；用途广；底片能提供检测记录	要求安全防护措施；价格贵；检测裂纹难度大，除非裂纹方向垂直射线底片	管道焊缝的穿孔、夹杂、气孔；铸造件的内部缺陷等
目视检测	表面特征如光洁度、划痕裂纹，或颜色；透明材料的应变；腐蚀	很方便，能够实现自动化	只能用于表面检测，检测表面开口缺陷，或者透明材料	纸、木头，或者金属的均匀、光洁表面的处理
涡流检测	由于材料的不同种类、裂纹、气孔、夹杂改变了电导率	容易实现自动化；花费一般	局限于导电材料；探测深度有限制	热交换管的薄壁和裂纹
液体渗透检测	由裂纹、小孔、接合缝，或者折叠产生的表面开口	价格不贵；易于使用；便携；对表面小裂纹的灵敏度高	缺陷必须在表面开口；对多孔材料和表面粗糙材料不起作用	涡轮机叶片表面裂纹和小孔；或者磨削裂纹
磁粉检测	表面或近表面的缺陷如裂纹、气孔、夹杂物，或材料几何形变引起的漏磁通	价格不贵或者花费一般；对表面和近表面的缺陷灵敏度高	局限于铁磁性材料；需要表面处理和检测后处理的；退磁处理	铁路轮子的裂纹；大型铸造件

无损检测的目的

由于人们对产品性能的期望值普遍增高，因此，确保产品的可靠性是必要的。无损检测在设计更轻、更坚固、更低成本、更可靠的零件方面起着非常重要的作用。

几乎每种无损检测方法都以一种或其他方式协助生产过程控制，从而确保制造商的直接利润。

无损检测可以应用于零件制造的每个阶段。原材料可以使用无损检测方法进行检查，可以接受或拒收。在锻造、焊接、热处理、磨制等生产过程中出现的缺陷也可以用无损检测方法进行检查。零部件在使用过程中发生的缺陷，如疲劳裂纹、腐蚀，也可以通过无损检测进行检查。

New Words

1. eddy   *n.* 涡流，漩涡
2. ultrasonic   *adj.* 超声的；*n.* 超声波
3. radiography   *n.* 射线照相术
4. acoustic   *adj.* 听觉的，有声的
5. optical   *adj.* 光学的，视力的
6. undercut   *n.* 咬边
7. overlap   *n.* 飞边，焊瘤
8. kit   *n.* 工具，器材，套，组
9. reinforcement   *n.* 加强法，（焊缝）加厚
10. capillary   *n.* 毛细管

Phrases and Expressions

1. nondestructive testing   无损检测
2. visual examination   肉眼观察
3. liquid penetrant   液体渗透
4. magnetic particle   磁粉
5. eddy current   涡流
6. ultrasonic   超声波

7. acoustic emission　声发射

8. thermal and optical　热和光

Some Common Abbreviations Used for Nondestructive Testing

ASTM　American Society for Testing and Materials　美国材料与试验协会

ASME　American Society of Mechanical Engineers　美国机械工程师协会

IIW　International Institute of Welding　国际焊接学会

ICNDT　International Committee for Nondestructive Testing　国际无损检测委员会

ISO　International Organization for Standardization　国际标准化组织

NDE　nondestructive evaluation　无损评价

It is a method to analyze material properties, and locate and characterize flaws within materials, fabricated parts and assemblies.

NDT　nondestuctive testing　无损检测

It is a method to locate and characterize material conditions and flaws, and be performed in a manner that does not affect the future usefulness of the object or material.

NDI　nondestructve inspection　无损检验

RT　radiographic testing　射线检测

It is a method that uses penetrating γ-or X-ray on materials and products to look for defects or examine internal or hidden features.

ICT　industrial computed tomography　工业计算机断层扫描

NMR　nuclear magnetic resonance　核磁共振

RA　radioactivity　放射性强度

IQI　image quality indicator　像质计

CR　computed radiography　计算机射线照相技术

DR　digital radiography　数字化射线照相技术

Gy Gray　戈瑞

UT　ultrasonic testing　超声检测

It is a method that uses sound waves of short wavelength and high frequency to detect flaws or measure material thickness.

AE　acoustic emission　声发射

T/R transmitter and receiver 发射器和接收器

dB decibel 分贝

Hz Hertz 赫兹

DAC distance amplitude correction 距离幅度校正

DGS distance gain size curve 距离幅度曲线

TOFD time of flight diffraction 衍射时差法

SNR signal to noise ratio 信噪比

MPI magnetic particle inspection 磁粉检验

MPT magnetic particle testing 磁粉检测

It is a method that can be used to find surface and near surface flaws in ferromagnetic materials by using the principle that magnetic lines of force (flux) will be distorted by the presence of a flaw in a manner that will reveal its presence.

MFL magnetic flux leakage 漏磁通

ET eddy current testing 涡流检测

It is an electromagnetic NDT method based on the process of inducing electrical currents into a conductive material by magnetic induction and observing the interaction between the currents and the material; in France it is known as the foucault currents method.

AC alternating current 交流电

DC direct current 直流电

A/D analog to digital 模拟/数字转换

LPI liquid penetrant inspection 渗透检验

It is a method that is used to reveal surface breaking flaws by bleedout of a coloured or fluorescent dye from the flaw.

PT penetrant testing 渗透检测

UV ultraviolet 紫外线

Chapter 8　The Weldability of Metals

Weldability is a much-used word, obviously intended to describe the ease or otherwise with which a metal may be welded. [①]Actually, there is no recognized method of classifying the weldability of the different metals—there are too many variables. However, it is possible to attempt a brief survey of those factors which affect the weldability (for the fusion welding processes) of the most commonly used metals, that is, there thermal characteristics, and their reaction to the application of welding heat.

Generally, the factors which decrease the weldability, or increase the fusion welding difficulties of metals are:

1. Very high or very low heat conductivity.

2. High degree of expansion when heated (high thermal expansion).

3. Low strength when hot (hot shortness).

4. Cold brittleness (cold shortness).

5. Tendency to oxidize readily when hot.

6. Tendency for the weld to harden by air cooling or the quench effect of surrounding cold metal. [②]

With a metal having high thermal conductivity e. g. copper, heat applied at the weld point is dissipated quickly into the remainder of the metal. On one hand, this necessitates a large volume, or a more intense source of welding heat, or preheating, in order to make up for the heat loss, hence the difficulty in spot welding copper. On the other hand, with a low thermal conductivity metal, e. g. stainless steel, the heat is concentrated in a narrow band along the joint, this may tend to overheating of the joint zone and restricts the area in which expansion can take place. It is, thus, often desirable to use copper bars under or at the sides of the joint to assist in heat dispersal.

Metals with a high coefficient of thermal expansion, e. g. aluminium and copper, expand and contract more for a given heat input than those with a low expansion coefficient, thus creating difficulties in counteracting distortion.

A metal which is hot short has low strength at welding temperatures, necessitating care to avoid contraction restraint and movement of workpieces during welding; small welds, such as tacks on cold metal, may crack when cooling. Cold shortness, indicates low strength at normal temperatures, e. g. cast iron; care is, therefore, necessary to minimize expansion at the weld point in order to avoid the creation of stresses at other points of the casting.

A metal which oxidizes readily at welding temperatures, e. g. aluminium, is difficult to weld due to the need to minimize atmospheric contact with the molten metal and to use fluxes for elimination or control of any oxide formed. This is one reason why the inert gas arc processes improve the weldability of stainless steel and aluminium: the weld zone is completely protected from oxidation and no flux is required.

A tendency to harden by rapid cooling necessitates the use of preheat to remove the quench effect of cold metal, and to reduce the cooling rate.

For the spot, projection and seam welding processes, the electrical conductivity of a metal is an additional factor affecting its weldability: high conductivity (or low resistance) metals—such as copper and aluminium—allow the current to pass across the joint with less generation of heat than would be the case with, say, steel, hence the need for exceptionally high currents for these metals.

Questions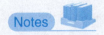

1. What is the weldability of metals?
2. What happens when welding stainless steel?
3. How many factors are there that influence the weldability of metals?
4. Why is the aluminium difficult to weld?

Notes

①Weldability is a much-used word, obviously intended to describe the ease or otherwise with which a metal may be welded.

焊接性是一个常用的名词，显然它是用来描述金属材料焊接的难易程度的。

句中 or otherwise 为插入语，可译为"相反"；describe the ease or otherwise 可译为"描述难易程度"；with which 引导定语从句，修饰 the ease or otherwise。

②Tendency for the weld to harden by air cooling or the quench effect of surrounding cold metal.

焊缝因空气冷却或周围冷金属的淬火效应而硬化的趋势。

句中...for the weld to harden by...为一带有逻辑主语的动词不定式短语作定语，修饰名词 tendency，可译为"焊缝由于……而变硬的倾向"。

参考译文

金属材料的焊接性

焊接性是一个常用的名词，显然它是用来描述金属材料焊接的难易程度的。实际上，衡量不同金属的焊接性没有固定的方法，因为影响金属材料焊接性的变量太多了。然而，对一些常用的金属材料，可以尝试观察影响其焊接性（熔化焊）的一些主要因素，如材料的热特性和它们对焊接热源的敏感程度。

通常情况下，可以降低金属材料的焊接性，即增加其熔焊难度的因素有：

（1）非常高或非常低的热传导性；

（2）在加热过程中变形很大（高的热胀系数）；

（3）受热时强度较低（热脆性）；

（4）冷脆性；

（5）加热时很容易氧化；

（6）焊缝因空气冷却或周围冷金属的淬火效应而硬化的趋势。

焊接导热性高的金属，如铜，施加于焊接处的热量迅速扩散到母材的其余部分。一方面为了弥补热量的散失，就需要一种大体积的，或者能量更集中的热源，或者是预热，因此点焊铜材的焊接性较差。另一方面，那些导热性低的金属，如不锈钢，热量集中在沿接头很窄的带状范围内，这可能导致接头区过热而且限制热膨胀。因此，通常将铜块放置在接头的下面或旁边以利于热量的散失。

在热输入相同时，热胀系数高的金属，如铝和铜，比热胀系数低的金属的膨胀和收缩量更大，因此，在抵抗变形方面难度更大。

具有热脆性的金属在焊接温度下强度较低，因此，在焊接期间需要采取措施防止收缩受阻和工件移动。较短的焊缝，如定位焊缝，由于在冷金属上施焊，在冷却过程中可能开裂。另一方面，具有冷脆性的金属，例如铸铁，在常温下强度就低，因此，需要采取措施尽量减少焊接区的膨胀，以免铸件的其他部位产生应力。

在焊接温度下极易氧化的金属，如铝，由于必须使大气与熔池金属尽量减少接触，很难焊接，因此就必须使用焊剂来消除形成的氧化物。这就是惰性气体保护可以改变不锈钢和铝的焊接性的一个原因。焊接区被保护起来，免受氧化，同时，不需要焊剂。

快速冷却条件下具有硬化趋势的金属在焊接时，需要采用预热来消除冷金属的淬火效应，同时可以降低冷却速度。

对于点焊、凸焊和缝焊过程，电极的导电性也是影响焊接性的另一个因素，高导电性（或低电阻）的金属，如铜和铝，在电流流过接头时产生的热量，要比钢产生的热量少，因此焊接这些金属需要特别大的电流。

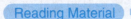

The Selection of Metals in Welding

Most of the metals in nature are used in industry, and about one-third of the metals are of interest in welding. A use can always be found for the particular combination of properties represented by any metal. Besides, the range of available metals has been greatly extended by alloying, and there is no limit to the number of alloys possible to be developed. From the metal iron, probably more than 25,000 alloys of steel are found to have been developed, although the uses of pure iron are very few in contrast to the vast range of purposes served by many steel alloys.

The basic characteristics that make the metals so very useful are their weldability, hardness, stiffness, and ductility (that is the property to be shaped easily). These characteristics of metals are of great importance to a welder. If he has some knowledge of them, he is certain to have his welding jobs done in a proper way.

There is no such thing as an ideal metal, although nickel may be considered to have come closest. For a particular application, the best selection is the metal that has the most favourable features. Probably the two most important characteristics to be considered are cost and mass. Mass particularly has been paid greater attention to in recent years owing to the development of modern industries. It is for this purpose that aluminium and beryllium are used to make aircraft and rockets.

Cost must be balanced against other characteristics of the metal. For example, it is

possible to replace beryllium by metals of only a hundredth the cost, but such substitutes cannot match beryllium's stiffness and strength, and the industries concerned may decide that the superiority of beryllium warrants the increased cost. For other applications, however, beryllium is not in such a favour, and therefore is not considered as a metal to be required.

Expansion in Welding

One of the greatest difficulties to be struggled with in welding is the control of expansion and contraction owing to differences in temperature of different parts of the piece welded. In many cases although distortion and crack have not taken place the expansion and contraction effects have produced serious internal strains. It is for this reason that many welded pieces which appear to have been successfully welded prove to be failures after being put back into use.

When metals are heated they expand in every direction; and conversely a decrease in temperature causes them to contract and the volume and linear dimensions decrease. In many practical cases we are only concerned with the increase in length. Increase in length is referred to as "linear" expansion, and increase in volume is called "cubical" expansion.

It is easy to calculate the amount of expansion to be expected from a rod of the metal of unit length when it is raised through one degree of temperature. This amount is called the "coefficient of linear expansion". Cubical expansion may be taken as three times the linear expansion.

New Words

1. weldability　　*n.* 可焊性，焊接性
2. survey　　*vt.* 观察
3. dissipate　　*v.* 消耗，消除
4. necessitate　　*vt.* 需要
5. dispersal　　*n.* 散开，散布
6. coefficient　　*n.* 系数
7. contract　　*v.* 收缩

8. counteract   *vt.* 抵抗，抵消

9. distortion   *n.* 变形，畸形

10. restraint   *n.* 抑制，约束

11. tack   *n.* 定位焊缝

12. crack   *n.* 裂纹

13. elimination   *n.* 消除

14. seam   *n.* 缝，接缝

Phrases and Expressions

1. much-used   常用的，经常被使用的

2. intend to   打算，想

3. make up for   弥补，补偿

4. assist in   参与，参加

5. due to   由于

Glossary of Terms

1. weldability   焊接性

2. heat conductivity   导热性

3. thermal expansion   热膨胀

4. hot shortness   热脆性

5. cold brittleness   冷脆性

6. cold shortness   冷脆性

7. expansion coefficient   膨胀系数

8. inert gas   惰性气体

9. expansion and contraction   膨胀和收缩

Glossary of Terms for Defects

1. weld undercut　咬边

A groove at the toe or root of a weld caused by melt.

2. gas porosity　气孔

Minute voids distributed in a casting that are caused by the release of gas during the solidification process.

3. crack　裂纹

A breach in material continuity in the form of a narrow planar separation.

4. hot crack　热裂纹

Cracks that are caused by differential contraction between a casting and its mold. These may be branched and scattered through the interior and on the surface of the casting.

5. brittle crack　脆性裂纹

The propagation of a crack that requires relatively little energy and results in little or no plastic deformation.

6. hydrogen embrittlement　氢脆

Low ductility caused by cracking of the interior of a component due to the evolution and precipitation of hydrogen gas.

7. corrosion fatigue　腐蚀裂纹

An acceleration of fatigue damage caused by a corrosive environment.

8. lamellar tearing　层间撕裂

The typical rupture of a material that is weakened by elongated slag inclusion. The crack surfaces have the appearance of wood fracture.

9. incomplete fusion (lack of fusion)　未熔合

Failure of weldment to fuse with the base material or the underlying or adjacent weld bead in a weld.

10. incomplete penetration (lack of penetration)　未焊透

Fusion into a joint that is not as full as dictated by design.

11. sag　夹渣

An aggregate of nonmetallic glasses that is found in primary metal production and in welding processes. Slag may be used as a covering to protect the underlying alloy from oxidation.

12. scratch　划痕

Small grooves in a surface created by the cutting or deformation from a particle or foreign protuberance moving on that surface.

13. crater　凹坑

A local depression in the surface of a component caused by excessive chip contact in machining or arc disturbance of a weldment.

14. heat-affected zone　热影响区

The portion of base material that was not melted during brazing, cutting, or welding whose microstructure and physical properties have been altered by the heat of the joining operation.

15. lap　搭接

A surface discontinuity appearing as a seam caused by the folding over of hot alloy fins, ears, or corners during forging, rolling, or other plastic forming without fusing the folds to the underlying material.

16. segregation　偏析

Nonuniform distribution of alloying elements, impurities, or phases in alloys.

17. casting shrinkage　铸造收缩

The reduction in component volume during casting. The reductions are caused by ①liquid shrinkage, which is the reduction of volume of the liquid as it cools to the liquidus temperature; ②solidification shrinkage, which is the total reduction of volume of the alloy through solidification; ③the shrinkage of the casting as it cools to room temperature.

18. cold shut　冷隔

A discontinuity on the surface of a casting caused by the impingement of melt (without fusion) within a part of a casting.

19. flake　白点

Short internal fissures in ferrous materials caused by stresses produced by evolution of hydrogen after hot working. Fractured surfaces containing flakes with bright and shiny surfaces.

20. inclusion 夹杂

Usually a solid foreign material that is encapsulated in an alloy. Inclusions comprised of compounds such as oxides, sulphides, or silicates are referred to as nonmetallic inclusions.

21. fold 折叠

Discontinuities composed of overlapping surface material.

22. lamination 分层

Separation or structural weakness, usually in plate that is aligned parallel to the surface of a component. This may be caused by the elongation during plastic forming of segregates that are the result of pipe, blisters, seams, and inclusions.

23. fatigue 疲劳

Progressive cracking leading to fracture that is caused by cyclic tensile load in the range of elastic stress, eventually initiating small cracks that sequentially and irreversibly enlarge under the action of fluctuating stress.

24. creep crack 蠕变裂纹

Cracking that is caused by linking of creep voids at the end of tertiary creep.

Chapter 9　Ultrasonic Testing

Ultrasonic testing（UT）uses a component to send a high-frequency vibration（beyond 20 kHz）and observes what happens when the beam hits a discontinuity or a change in density. The altered ultrasonic signal can be used to detect flaws within the material, to measure thickness from one side, and to characterize metallurgical structure.

Fig. 9 - 1 shows pulse reflection method for ultrasonic testing. The sending transducer（usually a piezoelectric crystal）transforms a voltage burst into ultrasonic vibrations. The transducer is coupled to the workpiece by a liquid medium, such as water. A receiving transducer converts the received ultrasonic wave into a corresponding electrical signal（Fig. 9-1（a））. With appropriate instrumentation, the same transducer alternately serves both functions in a so-called pitch-catch mode（Fig. 9-1（b））.

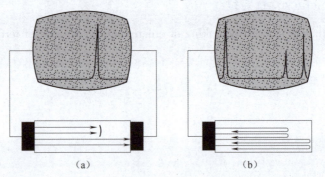

(a)　　　　　　　　　　　　(b)

Fig. 9-1　Pulse reflection method used for ultrasonic testing
(a) Ultrasonic method; (b) Pulse echo method

The signals are sent through the part and the time intervals that elapse between the initial pulse and the arrival of the various echoes are displayed on an oscilloscope screen.[①] A flaw is recognized by the relative position（on the time scale）and amplitude of the echo.

An angle transducer is often used at places where the normal beam transducer which contacts directly with the part surface can not reach. The ultrasonic beam can be used in an angular position by inserting a plastic wedge under the transducer（Fig. 9-2）. The beam bounces through the part until it uncovers a flaw, which causes an echo.

Fig. 9–2 Angel transducer method

Liquid immersion ultrasonic testing provides excellent coupling. Both the part and transducer are immersed, thus permitting reasonably fast scanning by using a traveling bridge arrangement over the part. [2] Numerical control (NC) and computers can be used to control and coordinate the motions for a quick display of defects.

Questions

1. How many kinds of transducers are described in this article?

2. Describe the applications of the ultrasonic testing.

3. What is the advantage of the immersion inspection?

Notes

①The signals are sent through the part and the time intervals that elapse between the initial pulse and the arrival of the various echoes are displayed on an oscilloscope screen.

被发射的信号经过零件和原始使脉冲与接收的各种回波的时间间隔都在示波屏上显示出来。

本句中 that 引导定语从句，作 the time intervals 的定语。

②Both the part and transducer are immersed, thus permitting reasonably fast scanning by using a traveling bridge arrangement over the work.

零件和探头都浸没在耦合液中，因此通过用一个放置在零件上的移动式电桥装置便可以实现恰当的快速扫描。

句中 both...and... 可译为"既……又……；……和……都"。

参考译文

超声波检测

超声波检测（UT）通过一种元件发射高频振动（超过 20 kHz）声波来观察当遇到不连续或密度变化时波束的变化。变化的超声波信号被用来检测材料内部缺陷、测厚、表征冶金组织状态等。

图 9-1 所示为脉冲反射法超声检测。发射探头（通常是一个压电晶片）把电能转换成超声振动能，探头通过水等液体介质与工件耦合。接收探头把接收的超声波转换成与之相对应的电信号（图 9-1（a））。用合适的仪器，相同的探头交替实现收-发模式的两种功能（图 9-1（b））。

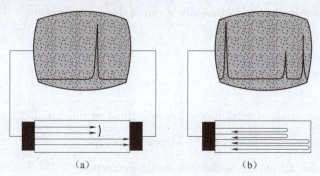

图 9-1　脉冲反射法超声检测

（a）超声波法；（b）脉冲回波法

被发射的信号经过零件和原始脉冲与接收的各种回波时间间隔都在示波屏上显示出来。（在时间轴上）缺陷可通过相对位置以及回波的波幅来检测。

斜探头常用于与零件表面直接接触的直探头探测不到的地方。在斜探头下方插入塑料楔，可以在角位置使用超声波束（图 9-2）。超声波束在零件中反射回来，直到缺陷被覆盖，从而产生回声。

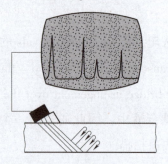

图 9-2　斜探头法

液浸超声波探伤提供了良好的耦合。零件和探头都浸没在耦合液中，因此通过用一个放置在零件上的移动式电桥装置便可合理地实现快速扫描。数字控制和计算机可用来控制和协调缺陷的快速显示。

Reading Material

Acoustic Emission Inspection

Engineering materials undergoing stress or plastic deformation emit sound. The acoustic emission is in the form of short bursts or trains of fast impulses in the ultrasonic range. These acoustic emissions can be related to the physical integrity of the material or structure in which they are generated, and the monitoring of these events permits detection and location of flaws as well as prediction of impending failure. The pulse rate and amplitude of acoustic emission bursts are usually very high compared to most natural or man-made noises, and therefore it is possible to isolate the significant signals by careful measurement of emission rates and amplitudes.

Measures or Defects　Cracking initiation and growth rate; Internal cracking in welds during cooling; Boiling or cavitation; Friction or wear.

Applications　Pressure vessels; Stressed structures; Turbine or gear boxes; Fracture mechanics research; Weldments.

Advantages　Remote and continuous surveillance; Permanent record; Dynamic (rather than static) detection of cracks; Portable.

Limitations　Transducers must be placed on part surface; highly ductile materials yield low amplitude emissions; Part must be stressed or operated; Test system noise needs to be filtered out.

Ultrasonic Testing Methods

Measures or Defects　Internal defects and variations: cracks, lack of fusion, porosity, inclusions, delaminations, lack of bond, texturing. Sound velocity; Corrosion ratio, elastic modulus.

Applications　Wrought metals, welds, brazed joints, adhesive-bonded joints, nonmetallics, in-service parts.

Advantages　Most sensitive to cracks; Test results known immediately;

Automating and permanent record capability; Portable; High penetration capability.

Limitations Couplant required; small, thin, complex parts may be difficult to check; Reference standards required; Trained operators for manual inspection.

Basic Knowledge of Ultrasonic Testing

At ultrasonic frequencies (above 20,000 Hz), sound energy can be formed into beams, similar to that of light, and thus can be scanned throughout a material, not unlike that of a flashlight used in a darkened room.

Types of Waves Waves that move in the same direction, or are parallel to their source are called longitudinal waves or compressional waves. Longitudinal sound waves are the easiest to produce and have the highest speed, however, it is possible to produce other types.

Reflection When ultrasonic waves encounter a discrete change in materials, as at the boundary of two dissimilar materials, they are usually partially reflected. If the incident waves are perpendicular to the material interface, the reflected waves are redirected back toward the source from which they came.

Refraction and Mode Conversion If the sound energy is partially transmitted beyond the interface, the transmitted wave may be ①refracted (bent), depending on the relative acoustic velocities of the respective media, and/or ②partially converted to a mode of propagation different from that of the incident wave.

Critical Angles The critical angles for the interface of two media with dissimilar acoustic wave velocities is the incident angle at which the refracted angle equals 90° (in accordance with Snell's law) and can only occur if the wave mode velocity in the second medium is greater than the wave velocity in the incident medium.

Attenuation Sound waves decrease in intensity as they travel away from their source, due to geometrical spreading, scattering, and absorption.

Wavelength and Defect Detection Sensitivity and resolution are two terms that are often used in ultrasonic inspection to describe a technique's ability to locate flaws.

Sensitivity is the ability to locate small discontinuities. Resolution is the ability of the system to locate discontinuities that are close together within the material or located near the workpiece surface.

Transducer Type Normal beam transducers are used to inspect relatively flat surfaces, and where near surface resolution is not critical.

Angle beam transducers are typically used to introduce a refracted shear wave into the test material (Fig. 9-3).

Calibration and Reference Standard Test block IIW (International Institute of Welding) is used extensively because this provides for the calibration of distance, sensitivity, resolution.

Contact & Immersion Techniques Contact testing is divided into three techniques, which are determined by the soundbeam wave mode desired: the straight-beam technique for transmitting longitudinal waves in the test workpiece, the angle-beam technique for generating shear waves, and the surface-wave technique for producing Rayleigh waves.

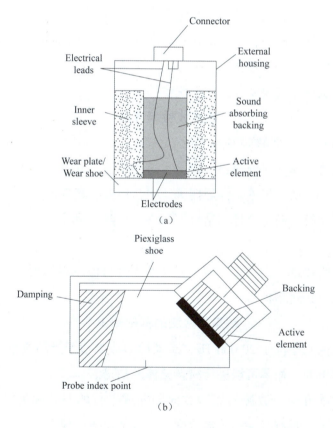

Fig. 9-3 Transducer type

(a) Normal beam transducer; (b) Angle beam transducer

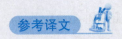 参考译文

声发射检测

工程材料在受到应力或塑性变形时会发出声音。声发射在超声波范围内呈短脉冲或快速脉冲的形式。这些声发射可能与产生它们的材料或结构的物理完整性有关，对这些进行监测可以检测和定位缺陷，并预测即将发生的故障。与大多数自然或人为噪声相比，声发射脉冲串的脉冲率和振幅通常非常高，因此可以通过仔细测量发射率和振幅来隔离区分重要信号。

测量或缺陷 裂纹萌生和扩展速率；焊缝冷却过程中的内部裂纹；沸腾或气蚀；摩擦或磨损。

应用 压力容器；受力结构；涡轮机或齿轮箱；断裂力学研究；焊接件。

优势 远程连续监控；永久记录；动态（而非静态）裂纹检测；便携式。

局限性 传感器必须放置在零件表面；高延展性材料产生低振幅发射；零件必须受压或服役；需要滤除测试系统中的噪声。

超声波检测方法

缺陷或测量 内部缺陷及其变化：裂纹、未熔合、孔隙、夹杂物、分层、未焊透、织构化；厚度或声速；泊松比、弹性模量。

应用 锻造金属、焊缝、钎焊接头、粘合接头、非金属、在役零件。

优势 对裂纹最敏感；检测结果即时性；自动化和永久记录能力；便携式；高穿透能力。

局限性 需要耦合剂；小、薄、复杂的零件可能难以检测；需要参考标准；操作人员必须要经过培训考核。

超声波检测的基础知识

当超声波的频率高于 20 000 Hz，声能即形成类似于光的波束，但它又不同于暗室中的手电筒，而是可以进行材料整体的扫描。

波型 传播方向与波源的振动方向同向或平行的声波称为纵波或压缩波。纵波最容易产生，具有最快的传播速度，还可转换成其他类型的声波。

反射 当超声波在材料内遇到不连续的变化，如两种相似材料的界面，通常会部分反射。如果入射波垂直于材料界面，反射波向与入射相反的方向传播。

折射与波型转换 如果部分声能透过界面，透射的超声波会①发生折射，折射的程度取决于两种介质的声速；②部分声波转换成其他波型。

临界角　具有不同声速异质界面的临界角是指当折射角达到 90°时对应的入射角（依据 Snell 定律），只有当第二种介质中声波的速度大于入射介质的声速才会出现临界角。

衰减　由于扩散、散射和吸收，声波在传播过程中随着传播距离的增加声波强度减弱的现象。

波长与缺陷检测　灵敏度和分辨率是超声检测中用来描述探测和定位缺陷能力的两个术语。

灵敏度是探测不连续性的能力；分辨率表示检测系统将材料中相距很近或位于靠近工件表面的不连续性分辨开的能力。

探头的种类　直探头用来检测具有表面相对平坦、对近表面分辨率要求不高的工件。

斜探头主要用于在被检材料中产生的折射横波（图 9-3）。

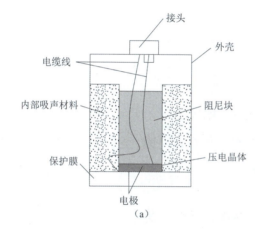

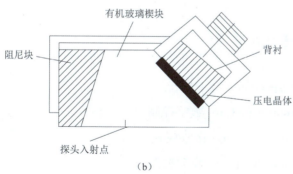

图 9-3　探头类型

(a) 直探头；(b) 斜探头

校验试块和参考试块　国际焊接学会（IIW）试块得以广泛使用是因为它可

以进行距离、灵敏度、分辨率的校准。

　　接触法和液浸法　接触法基于所需波型可以分为三种技术：在被检工件中传播纵波的直束检测技术，产生横波的斜束检测技术、产生瑞利波的表面波检测技术。

 New Words

1. component　*n.* 零件
2. frequency　*n.* 频率
3. vibration　*n.* 振动，振荡
4. characterize　*v.* 表征，表示……的特征
5. piezoelectric　*n.* 压电的
6. pitch　*v.* 发射
7. attenuate　*v.* 使变细，减少，衰减
8. elapse　*vi. & n.*（时间）过去，经过
9. oscilloscope　*n.* 示波器
10. amplitude　*n.* 振幅，幅度
11. immersion　*n.* 浸液，浸水，浸油
12. couple　*v.* 耦合
13. wedge　*n.* 楔块
14. bounce　*v.* 使（多次）反射，发射

 Phrases and Expressions

1. immersion inspection　液浸探伤
2. numerical control（NC）　数字控制
3. pitch-catch mode　发射-接收模式
4. high-frequency vibration　高频振荡
5. ultrasonic signal　超声信号
6. sending transducer　发射探头
7. ultrasonic wave　超声波
8. oscilloscope screen　示波器屏幕

Glossary of Terms for Ultrasonic Testing

1. acoustic impedance 声阻抗

The propagation of waves in solids depends on the resistance of the atoms (or particles) to vibrate, that is $Z=\rho C$. It can be shown that Z is given by the product of sound velocity (C) and density of a material (ρ).

2. transducer 探头

A device that converts energy from one form to another. An electroacoustical transducer converts electrical energy into acoustic energy and vice versa.

3. angle transducer 斜探头

Designed to project ultrasonic waves at an angle to the surface to generate shear, surface, or other types of wave mode conversion.

4. piezoelectric effect 压电效应

Observed in certain crystals with low symmetry atomic structures where a potential difference develops across opposite faces when the solid is subjected to a stress.

5. ultrasound (ultrasonic radiation) 超声波

A mechanical wave at a frequency above 20 kHz.

6. attenuation (absorption) 衰减

The decrease in ultrasonic intensity with distance expressed in decibel (dB) per unit length or nepers (Np) per unit length. (1 dB = 8.686 Np/cm)

7. A scan A-扫描

A plot of signal amplitude against time that can be related to distance in a workpiece.

8. pulse echo method 脉冲回波法

The presence and position of a discontinuity are given by the echo amplitude and time.

9. compression wave (longitudinal wave) 纵波

Composed of a series of alternate surfaces of cornpressions and rarefactions traveling perpendicular to these surfaces. Particle motion is in the direction of travel.

10. transverse wave 横波

The particle displacement at each point in a material is perpendicular to the direction of wave propagation.

11. Rayleigh wave（surface wave） 瑞利波（表面波）

Surface wave in which the particle motion is elliptical and the effective penetration is of the order of one wavelength.

12. Lamb wave 兰姆波

A mode of propagation in which the two parallel boundary surfaces of the plate or the wall of a tube establish the mode of propagation.

13. mode conversion 波型转换

The conversion of the mode of ultrasonic wave propagation, which can occur at an interface between materials of different acoustic impedances.

14. refraction 折射

The change in the direction of a beam of radiation on passing through an interface.

15. critical angle 临界角

The angle of incidence beyond which a particular refracted wave is not transmitted.

16. dead zone 盲区

Directly in front of an ultrasonic transducer where no reflections can be detected due to the length of the initial pulse, or to amplify overload resulting from the initial pulse.

17. near field（Fresnel zone） 近场

The region of the ultrasonic beam adjacent to the transducer which is dominated by interference effects.

18. far field（Fraunhofer） 远场

Extends beyond the limit of the near field where the ultrasonic beam amplitude on the beam axis decreases exponentially with distance.

19. couplant 耦合剂

A substance used between the transducer probe and the surface to improve transmission of ultrasonic energy.

20. distance amplitude correction（DAC）curve 距离波幅曲线（DAC）曲线

The curve of amplitude responses obtained from reflectors of equal area situated at different depths within an object.

21. immersion testing 液浸法

Transducers and workpiece are immersed in a liquid bath, which act as acoustic coupling.

Chapter 10   Radiographic Testing

Radiographhic testing （RT） is essentially a shadow pattern created when certain types of radiation penetrate an object and are differentially absorbed depending on variations of thickness， density， or chemical composition of the material. [1] A schematic of a typical radiographic testing system is shown in Fig. 10 - 1. The shadowgraph is commonly registered on a photographic film to provide a permanent record. Other methods of registering the image include fluoroscopy， xerography， and closed-circuit television scanning.

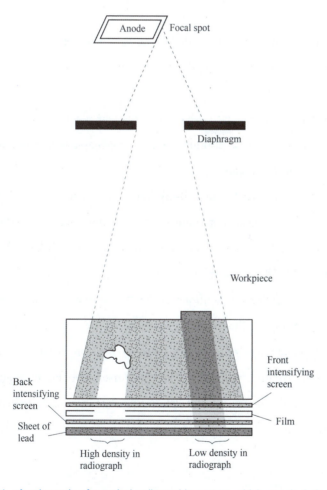

Fig. 10–1   A schematic of a typical radiographic system which may include the use of a diaphragm and screens to reduce scatter effect and enhance the final image

Three types of penetrating radiation are presently used for industrial radiography: X-rays, gamma (γ-) rays, and neutron beams.

X-Rays X-rays are a form of electromagnetic radiation similar to that of light, heat, or radio waves. A distinguishing feature of X-rays is their extremely short wavelength only about 1/10,000 that of light and it is this characteristic that enables X-rays to penetrate materials that absorb or reflect ordinary light. The X-rays are generated when electrons, traveling at high speeds, collide with matter. They are generally produced in an evacuated X-ray tube, which contains a heated filament and a target. A high voltage across the filament (cathode) causes it to emit electrons, which are then driven to the target (anode) where the sudden deceleration results in X-rays. [②] The higher the tube voltage, the greater the energy and penetrating power of the X-rays.

γ- Rays γ-rays are emitted by disintegrating nuclei of radioactive substances. In industrial radiography, artificially produced radioactive isotopes, such as cobalt 60, are used almost exclusively.

Neutron Rays Neutrons are derived from nuclear reactors, nuclear accelerators, or radioactive isotopes. For most applications, it is necessary to moderate the neutron energy and to collimate the beam.

In general, the X-ray and γ-ray absorption of the material tested depends on the thickness, density, and most importantly, the atomic nature of the material. For example, lead is about 1.5 times as dense as steel, but at 220 kV, 2.54 mm of lead absorbs as much as 30.48 mm of steel.

Although the absorption of neutrons by matter is comparable with that of high-energy X-rays or γ-rays, the relative absorption by various elements is quite different. There is no obvious relationship between neutron absorption and the atomic number of a material. Neutron radiography has been useful in obtaining images of organic material within a metallic housing, and also in the detection of hydrogen embrittlement in engineering materials.

Questions

1. Presently, what types of penetrating are used for industrial radiography?

2. How are the X-rays generated?

3. What are the differences among the generation of X-rays with γ-rays and neutron beams?

Notes

①Radiography testing is essentially a shadow pattern created when certain types of radiation penetrate an object and are differentially absorbed depending on variations of thickness, density, or chemical composition of the material.

射线照相（RT）实质上是当一定类型的射线穿过物体时，产生的一种影像图案。物质的厚度、密度，或化学成分的不同，其对射线吸收程度就不同，从而产生阴影图案。

句中 when certain types of radiation penetrate an object 为一个时间状语从句。

②A high voltage across the filament (cathode) causes it to emit electrons, which are then driven to the target (anode) where the sudden deceleration results in X-rays.

高电压通过阴极使它发射电子，电子在高压下运动直接打到阳极靶上并突然停止，从而产生 X 射线。

句中由 which 引导的定语从句修饰 electrons，where 引导的定语从句则修饰 the target (anode)。

参考译文

射线检测

射线照相（RT）实质上是当一定类型的射线穿过物体时，产生的一种影像图案，物质的厚度、密度，或化学成分的不同，其对射线吸收程度也不同，从而产生阴影图案。图 10-1 所示是一种典型的射线照相检测系统示意图。影像通常被记录在相纸上而提供永久的记录，其他记录影像的方法包括荧光法、静电复印术和闭路电视扫描等。

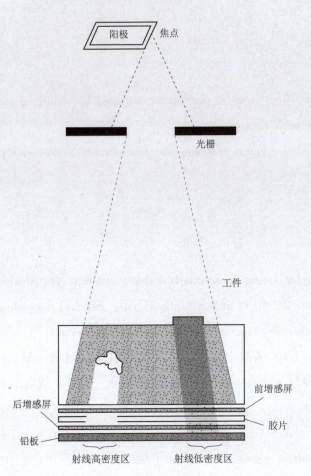

图 10-1　典型射线照相检测示意图（包含使用光栅和增感屏来减少散射效应并增强最终影像）

在工业射线照相中，目前有三种类型的穿透射线可用：X 射线、γ 射线和中子射线等。

X 射线　X 射线是一种与光、热、无线电波相似的电磁射线形式。X 射线明显的一个特征是波长极短只有大约 1/10 000 的光波波长，它的这种特性能让 X 射线穿透材料，这些材料会吸收或反射普通光。当高速传播的电子撞击物质时产生 X 射线。它们通常在一个真空管中产生，这个管包括一个热的阴极丝（阳极）和一个（阳极）靶。通过阴极的高电压可使阴极丝发射电子，然后电子（在高压下）运动到（阳极）靶上，突然减速从而产生 X 射线。管的电压越高，X 射线的能量和穿透力就越强。

γ 射线　γ 射线是放射性物质核衰变时放出的。在工业射线照相时，人工放射性同位素，如钴 60 几乎是专用的放射性同位素。

中子射线　中子是在核反应中核加速或由放射性同位素产生的。对于大多数的装置而言，使中子能量适中和控制中子束是必要的。

通常，被检材料所吸收的X射线、γ射线取决于其厚度、密度，而且最重要的是材料的原子量。比如，铅的密度是钢的1.5倍，但220 kV、2.54 mm厚的铅板吸收射线的量相当于30.48 mm厚的钢板吸收的量。

虽然物质对中子的吸收量与高能X射线或γ射线的量相当，但不同元素吸收的相对量是完全不同的。材料对中子的吸收与原子量没有明显的关系。中子照相已用于金属制的房屋中有机材料的成像，也用于工程材料中氢脆的检测。

Reading Material

Inspection by Neutron Radiography

Neutron radiography is a form of nondestructive inspection that uses a specific type of particulate radiation, called neutrons, to form a radiographic image of a test workpiece. The geometric principles of shadow formation, variation of attenuation with test workpiece thickness, and many other factors that govern the exposure and processing of a neutron radiograph are similar to those for radiography using X-rays or γ-rays.

Neutrons are subatomic particles that are characterized by relatively large mass and a neutral electric charge. The attenuation of neutrons differs from the attenuation of X-rays in that the processes of attenuation are nuclear rather than ones that depend on interaction with electron shells surrounding the nucleus.

Using neutrons, it is possible to radiographically detect certain isotopes—for instance, certain isotopes of hydrogen, cadmium or uranium. Some neutron-image-detection methods are insensitive to γ-rays or X-rays, and can be used to inspect radioactive materials such as reactor fuel elements. The high attenuation of hydrogen, in particular, opens many application possibilities, including inspection of assemblies for detection of adhesives, explosives, lubricants, water, corrosion, plastic or rubber.

Radiography Testing Methods

Radiography（*X-Rays Film and Image Tubes*）

Measures or Defects　Internal defects and variations: porosity, cracks, lack of fusion, inclusions, geometry variations, corrosion, density variations.

Applications Castings; electrical assemblies; weldments; small, thin, complex wrought products; nonmetallics; solid propellant rocket motors.

Advantages Permanent records; film; adjustable energy levels; high sensitivity to density changes; no couplant required; geometry variations do not affect direction of the X-ray beam.

Limitations High initial costs; orientation of linear defects in part may not be favorable; radiation hazard; depth of defect not indicated; sensitivity decreases with increase in scattered radiation.

Radiography（γ-Rays）

Measures or Defects Internal defects and variations: porosity, cracks, lack of fusion, inclusions, geometry variations, corrosion.

Applications Usually in places where X-ray machines are not suitable because source cannot be placed in part with small openings and/or power source not available.

Advantages Low initial cost, permanent records; film, small sources can be placed in parts with small openings, portable; low contrast.

Limitations One energy level per source, source decay, radiation hazard, trained operators needed, lower image resolution, cost related to energy range.

Basic Knowledge of Radiographic Testing

X-rays and γ-rays possess the capability of penetrating materials, even those that are opaque to light. In passing through the matter, some of these rays are absorbed.

The amount of absorption at any point is dependent upon the thickness and density of the matter at that point; therefore, the intensity of the rays emerging from the matter varies. When this variation is detected and recorded, usually on film, a means of seeing within the material is available.

Characteristics of X- and γ-Rays X-rays and γ-rays are electromagnetic radiation of exactly the same nature as light, but of much shorter wavelength.

As they pass through matter they are scattered and absorbed and the degree of penetration depends on the kind of matter and the energy of the rays.

Image Consideration Radiographic sensitivity is dependant on the combined effects of two independent sets of variable. One set of variables affect the contrast and the other set of variable affect the definition of the image.

X-Ray Generators   The tube cathode (filament) is heated with a low-voltage current of a few amps. The filament heats up and the electrons in the wire become loosely held. A large electrical potential is created between the cathode and the anode by the high-voltage generator. Electrons that break free of the cathode are strongly attracted to the anode target. The stream of electrons between the cathode and the anode is the tube current. Fig. 10-2 shows the components of X-ray generator.

Selection of Energy   For X-rays below 150 kV the choice of correct kilovoltage is important because the attenuation coefficient varies rapidly. From 200 to 400 kV, only a considerable difference on the order of 30-40 kV, will make a significant difference in sensitivity.

X-Ray Exposure Charts   An exposure chart usually applies only to a single set of conditions, determined by ①the X-ray machine used; ②a certain source-to-film distance; ③a particular film type; ④processing conditions used; ⑤the film density on which the chart is based; ⑥the type of screens (if any) that are used; and ⑦the workpiece tested.

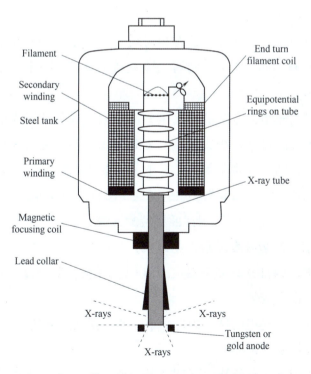

Fig. 10-2   X-ray generator

Exposure Factor   The exposure factor is a quantity that combines milliamperage

(X-rays) or source strength (γ-rays), time and distance.

Film Processing Processing film is a strict science governed by rigid rules of chemical concentration, temperature, time, and physical movement. Whether processing is done by hand or automatically by machine, excellent radiographs require a high degree of consistency and quality control.

Exposure Control Techniques The concepts of time, distance, and shielding can be used to control the amount of exposure received by personnel working with sources of radiation.

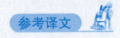

中子射线照相检测

中子射线照相是一种无损检测形式，它使用一种特定类型的辐射颗粒，称为中子，从工件的射线照相中提取。中子射线阴影形成的几何原理、衰减随工件厚度的变化情况以及控制中子射线照片曝光和处理的许多其他因素，与使用 X 射线或 γ 射线的射线照相相似。

中子是亚原子粒子，其特征是有相对较大的质量和带中性电荷。中子的衰减与 X 射线的衰减不同，其衰减过程主要是核的衰减，而不是依赖于核周围电子壳的相互作用。

使用中子射线照相，可以检测出某些同位素，如氢、镉或铀的同位素。一些中子射线检测方法对 γ 射线或 X 射线不敏感，可用于检查像反应堆燃料元件的放射性材料。特别是氢的高衰减，还开发了许多可能应用，包括检测集成件，以检测黏合剂、爆炸物、润滑剂、水、腐蚀、塑料或橡胶等。

射线照相检测方法

射线照相术（X 射线胶片和 X 射线机）

测量或缺陷 内部缺陷及其变化：孔隙、裂纹、未熔合、夹杂物、几何形状变化、腐蚀、密度变化。

应用 铸件，电气集成件，焊件，小、薄、复杂的锻造产品，非金属材料，固体推进剂的火箭发动机。

优势 永久记录；有胶片；可调节的能级；对密度变化的高灵敏度；不需要耦合剂；几何形状变化不会影响 X 射线束的方向。

局限性 初始成本高；零件中线性缺陷的取向可能不利；有辐射危害；缺

陷的深度不可获得；灵敏度会随散射辐射的增加而降低。

射线照相术（γ射线）

测量或缺陷　内部缺陷及其变化：孔隙、裂纹、未熔合、夹杂物、几何形状变化、腐蚀。

应用　通常用在 X 射线机不适合的地方，因为射线源不能放置在小开口或电源不可用的地方。

优势　初始成本低，永久记录；有胶片，小的射线源可以放置在具有小开口的工件中，便携式；具有低对比度。

局限性　每个源只有一个能级，源会衰变，有辐射危害、需要训练有素的操作人员，影像分辨率较低，成本与能量范围相关。

射线检测基础知识

X 射线和 γ 射线具有穿透物质的能力，甚至能穿透不透明的物质。在穿透物质的时候，部分射线会被吸收。

在任何位置吸收的量取决于该处物质的厚度与密度；因此透过物质的射线的强度是随着位置的变化而变化的。当在胶片上检测并记录到这种变化时，就可以看到材料的内部了。

X 射线和 γ 射线的特性　X 射线和 γ 射线属于电磁波，具有与可见光一样的特性，只是波长要短很多。

当穿透物质时，射线被吸收和散射，射线穿透物质的能力取决于物质的种类以及射线的能量。

影像　射线检测的灵敏度取决于两组独立变量共同作用的结果。一组变量影响影像的对比度，另一组变量影响影像的清晰度。

X 射线发生器　阴极管（阴极灯丝）用几安培的低电压电流加热，加热后灯丝处的电子就会活跃起来。高压发生器在阴极管与阳极之间形成高压差，挣脱阴极的电子被阳极靶强烈吸引过去。阴极和阳极之间形成的电子流即管电流。图 10-2 所示为 X 射线发生器的组成部分。

能量的选择　低于 150 kV 的 X 射线的电压选择很重要，因为衰减系数的变化很大。从 200~400 kV，仅当管电压的差异大到 30~40 kV，其灵敏度的差异才显著。

X 射线曝光曲线　曝光曲线通常只在一定条件下适用：①X 射线机；②源到胶片的距离；③胶片类型；④暗室处理技术；⑤胶片密度；⑥感光屏类型；⑦被检工件材料。

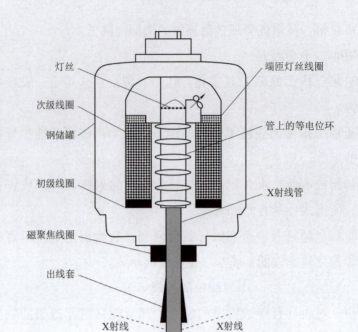

图 10-2　X 射线发生器

曝光因子　曝光因子给出的是 X 射线或 γ 射线的活度、曝光时间、焦距的关系。

暗室处理　胶片处理是射线照相检测的重要技术环节，受化学试剂浓度、温度、时间以及操作的严格控制。无论是手工处理还是自动处理，只有高度的一致性和质量控制才能获得优质的底片。

辐射控制方法　工作人员进行放射性工作时，可以利用时间、距离和屏蔽物来控制接收的辐射剂量。

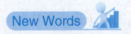

1. shadowgraph　*n.* 放射线透视照相

2. xerography　*n.* 静电印刷术

3. evacuate　*v.* 抽真空

4. filament　*n.* 阴极丝

5. cathode　*n.* 阴极

6. anode　*n.* 阳极

7. decelerate　*v.* 减速，降低，降速

8. disintegrate　*n.* 衰变，分解

9. cobalt　*n.* 钴

10. isotope　*n.* 同位素

11. collimate　*vt.* 校对，使成直线

12. hydrogen　*n.* 氢

Phrases and Expressions

1. close-circuit television scanning　闭路电视扫描

2. X-ray　X 射线

3. gamma ray　伽马射线

4. neutron beam　中子束

5. industrial radiography　工业用射线

6. short wavelength　短波长

7. almost exclusively　几乎完全，全部

8. comparable with（to）　可比较的，比得上的

9. metallic housing　金属结构

10. hydrogen embrittlement　氢脆

Glossary of Terms for Radiograph

1. radiograph　射线照相，X 射线照片

A permanent, visible image on a recording medium produced by penetrating radiation passing through the material being tested.

2. exposure, radiographic exposure　射线曝光量

The subjection of a recording medium to radiation for the purpose of producing a latent image. Radiographic exposure is commonly expressed in terms of milliampere-seconds or millicurie hours for a known source-to-film distance.

3. exposure range（latitude）　曝光量范围，曝光宽容度

The range of exposures over which a film can be employed usefully.

4. focal spot　焦点

For X-ray generators, that area of the anode (target) of an X-ray tube that emits X-ray when bombarded with electrons.

5. absorbed dose　吸收剂量

The energy imparted by ionizing radiation per unit mass of irradiated material; measured by "rad" where 1 rad = 0.01 J/kg. The SI unit of absorbed dose is Gray (Gy) where 1 Gy = 1 J/kg.

6. attenuation coefficient　衰减系数

Related to the rate of change in the intensity of a beam of radiation as it passes through matter.

7. back scattered radiation　背散射

Radiation which is scattered more than 90° with respect to the incident beam, that is, backward in the general direction of the radiation source.

8. Compton scattering　康普顿散射

When a photon collides with an electron it may not lose all its energy, and a lower energy photon will then be emitted from the atom at an angle to the incident photon path.

9. densitometer　光学密度计

A device for measuring the optical density of radiograph film.

10. density gradient　密度梯度

The slope of the curve of density against log exposure for a film.

11. equivalent image quality indicator (IQI) sensitivity　等效像质计灵敏度

That thickness of IQI expressed as a percentage of the section thickness radiologically examined in which a 2T hole or 2% wire size equivalent would be visible under the same radiological conditions.

12. contrast sensitivity　对比度灵敏度

A measure of the minimum percentage change in an object which produces a perceptible density/brightness change in the radiological image. Contrast stretch is a function that operates on the greyscale values in an image to increase or decrease image contrast.

13. radiographic contrast　射线底片对比度，射线照相等效系数

The difference in density between an image and its immediate surroundings on a radiograph.

14. density (film)　胶片密度

A quantitative measure of film blackening when light is transmitted or reflected.

15. fog　灰雾度

A term used to denote any increase in density of a processed photographic emulsion caused by radiation from sources other than intentional exposure to the primary beam; e. g. film may be exposed to scatter radiation, or accidental exposure may occur if stored film is not protected from radiation.

16. film contrast　底片对比度

The slope of the characteristic curve of a photographic material that is related to the density difference resulting from a given exposure difference.

Chapter 11　Magnetic Particle Testing

Magnetic particle testing (MPT) is based on the principle that ferromagnetic materials. When magnetized, the workpiece will have distorted magnetic fields in which there are material flaws and that these anomalies can be clearly shown with the application of magnetic particles.

The magnetic field can be set up by passing an electric current through all or a portion of the workpiece. The current may be passed through the workpiece or through a conductor in close proximity to the workpiece. The following figure shows magnetic particle testing principle.

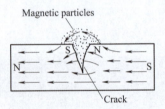

Magnetic particle testing

To be effective, the direction of the induced field should be almost perpendicular to the expected flaw.

Either AC or DC can be used to generate the magnetic field. Magnetization is better with AC for surface discontinuities while DC is used to locate subsurface discontinuities or nonmetallic inclusions.

Magnetic Particle Application　The magnetic particles can be applied either when the current is applied, which is the continuous technique, or after the current has been shut off, which is the residual technique. The sensitivity of the residual technique is lower but there is less chance of false indications produced by current leaks.

After the particles have been sprayed or sprinkled on the surface, the excess is gently removed by blowing or sweeping—leaving only the magnetic pattern.

The magnetic particles are available in several colors or treated with fluorescent material for observation under ultraviolet light.

Questions

1. What is the principle of magnetic particle testing?

2. Describe the applications of magnetic particle testing.

3. What is the difference between the residual technique and the continuous technique?

Notes

①When magnetized, the workpiece will have distorted magnetic fields in which there are material flaws and that these anomalies can be clearly shown with the application of magnetic particles.

当工件被磁化后，在工件材料中有缺陷部位的磁场将会扭曲，而且这种异常现象在施加磁粉后就能够很清楚地显示出来。

在本句中，由 in which 引导的定语从句修饰 magnetic fields。

②The sensitivity of the residual technique is lower but there is less chance of false indications produced by current leaks.

剩磁法的灵敏度（比连续法的灵敏度）低，但是它因电流泄漏而产生虚假显示的概率较小。

参考译文

磁粉检测

磁粉检测是基于铁磁性材料原理。当工件被磁化后，在工件材料中有缺陷部位的磁场将会扭曲，而且这种异常现象在施加磁粉后就能够很清楚地显示出来。

当电流通过整个工件或工件的一部分时就会产生磁场。电流可以通过工件或通过紧靠着工件的导体。下图所示为磁粉检测原理。

为了有效检测出缺陷，电磁感应场的方向应该几乎与预判的缺陷方向垂直。

无论是交流电还是直流电都可以产生磁场。对于表面缺陷，交流磁化的效果更好，而直流磁化主要用于检测近表面的缺陷或非金属夹杂物。

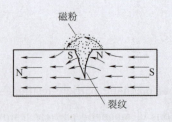

磁粉检测

磁粉检测的方法 磁粉可以在通电时施加，这是连续法，也可以在通电结束后再施加，这就是剩磁法。剩磁法的灵敏度（比连续法的灵敏度）低，但是它因电流泄漏而产生虚假显示的概率较小。

在磁粉喷涂或撒在被检工件表面后，多余的磁粉可以通过吹扫或清扫轻轻去除，只留下缺陷的磁粉图案。

磁粉也可以添加一些颜色，或者用荧光材料处理，以便在紫外光下进行观察。

 Reading Material

Basic Knowledge of Magnetic Particle Testing

Magnetic testing（MT）, also referred to as magnetic particle testing, is a relatively simple test method that can be applied to a variety of product forms such as castings, forgings, and welding.

Magnetic testing consists of magnetization of the workpiece, application of the particles, and interpretation of the patterns formed by the particles as they are attracted by magnetic leakage fields.

Flux Leakage Field The magnetic field spreads out when it encounter the small air gap created by the crack because the air cannot support as much magnetic field per unit volume as the magnet can.

If iron particles are sprinkled on a cracked magnet, the particles will be attracted to and cluster at the poles at the edges of the crack.

Stationary Unit Stationary magnetic particle testing units are usually large, not easily moved, are hard wired to a commercial source of electricity and have a wet bath built into them.

Most equipment has a thermal circuit breaker which interrupts the operation if an

overload occurs and, after sufficient cooling time, restores it again.

Electromagnetic Yoke Most yokes are also equipped with AC, which can be used for either demagnetization or AC inspection with dry particle techniques.

Some yokes have articulated legs that can be adjusted in two places in order to change the distance between the legs and thus vary the flux density.

Mediums To achieve the required test sensitivity, the degree of particle concentration in the bath must be correct—too light concentration leads to very light indications of discontinuities; too heavy concentration results in too much overall surface coverage, which may mask or cause incorrect interpretation of discontinuity indications.

Ultraviolet Light These lights are high pressure mercury vapor lamps that provide high intensity light with a fairly wide spectral range that extends through the ultraviolet and visible wavelengths and into the infrared region.

Magnetizing Currents The advantages of using AC including: best suited for locating surface discontinuities because of skin effect; the voltage can be stepped up or down; the reversal of magnetic fields due to the alternating current makes the magnetic particles more mobile to collect at leakage fields. The disadvantage is that AC has less penetration than DC.

Testing Techniques The first, based on whether or not the magnetizing force is maintained during application of the medium, includes the residual method and the continuous method.

Continuous and Residual Methods In the continuous method, the application of the medium is conducted simultaneously with the magnetizing operation, i. e. the medium is in contact with the test workpiece while current is being applied.

Circular and Longitudinal Methods Current is passed through the workpiece and a circular magnetic field is established in and around the workpiece. Care must be taken to ensure that good electrical contact is established and maintained between the test equipment and the test component.

Demagnetizing The basis for all demagnetization methods is the subjecting of the magnetized workpiece to the influence of a continuously reversing magnetic field that gradually reduces in strength causing a corresponding reversal and reduction of the field in the workpiece.

Testing Sequence The following steps must be completed for a test in a stationary unit：

(1) Workpiece preparation；

(2) Setting up the equipment；

(3) Application of the particles；

(4) Application of the current；

(5) Inspection of the workpiece；

(6) Demagnetization；

(7) Postcleaning of the workpiece；

(8) Documentation of the test.

Interpretation and Evaluation of Indications. A nonrelevant indication is an indication caused by something that does not interfere with the use of a workpiece, whereas a relevant indication is caused by something that may interfere with the use of the workpiece.

磁粉检测的基础知识

磁粉检测（MT）也可以称为磁粉检验（MPI），是一种相对简单的检测方法，适用于铸件、锻件和焊接件等各种产品的检测。

磁粉检测的过程包括工件磁化、磁粉施加，以及漏磁场吸附磁粉形成的磁痕的解释。

漏磁场 由于空气传导磁场的能力比磁铁差，所以，当磁场遇到裂纹内的空气层时就会溢出工件。

如果铁粉洒到有裂纹的磁体，磁粉会被吸引并聚集在裂纹形成的磁极周围。

固定式探伤机 固定式磁粉探伤机体积大，不易移动，与商用电源连接，内置储液槽。

大部分的设备都具备热断路器，当发生过载的时候可以中断操作，经过一定的冷却时间可再重新启动。

电磁轭 大部分磁轭都配备交流电源，既可用于退磁也可以交流电的干法检验。

部分磁轭的两个磁极是可调的关节，可以在两个位置上调节以改变磁极间的距离，进而调节磁通密度。

磁介质　为了获得所需的检测灵敏度，磁悬液的浓度必须合适——浓度太低会导致显示不清晰，浓度太高会导致背景显示过重，这样会掩盖不连续性的显示，或者造成不连续性显示的错误解释。

紫外灯　紫外线灯是高压汞蒸气灯，它可以提供具有相当宽光谱范围的高强度光，可通过紫外线和可见光波长延伸到红外区域。

磁化电流　交流电的优势包括：由于集肤效应，最适合表面不连续性的检测；电压是交变的；由于交变电流产生的交变磁场使得磁粉在漏磁场处具有更好的流动性。交流电的不足使得渗透性不如直流电。

检测技术　首先，根据施加磁介质的过程中是否保持磁化，磁粉检测技术包括连续法和剩磁法。

连续法和剩磁法　对于连续法，施加磁介质需要与磁化过程保持同步，即通电时需要将磁介质施加到工件上。

周向磁化法和纵向磁化法　工件通电，在工件内部以及周围形成一个环形磁场。必须小心确保磁化装置与被检工件之间建立并保持良好的电接触。

退磁　所有退磁方法的基础是将被磁化的工件置于持续反转且强度逐渐减小的磁场中，导致工件中的磁场也反转并减小。

检测步骤　在固定式磁粉探伤机上要完成以下过程：

（1）预处理；

（2）安装设备；

（3）磁介质施加；

（4）通电；

（5）观察工件；

（6）退磁；

（7）后清洗工件；

（8）检测文档。

显示的解释和评价　不相关显示是由并不会妨碍工件使用的某个因素引起的，而相关显示是由妨碍工件使用的某个因素引起的。

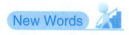New Words

1. anomaly　*n.* 反常情况，异形物

2. portion   *n.* 部分，区段

3. proximity   *n.* 接近，贴近

4. subsurface   *adj.* 表面下的，地下的

5. residual   *adj.* 剩余的，残留的

6. sprinkle   *v.* 撒，洒，喷

7. ultraviolet   *adj.* 紫外线的，产生紫外线的

Phrases and Expressions

1. magnetic particle   磁粉

2. magnetic particle testing   磁粉检测

3. in close proximity to   紧靠着

4. AC or DC   交流或直流

5. sprayed or sprinkled   喷撒

6. ultraviolet light   紫外线光

Glossary of Terms for Magnetic Particle

1. ferromagnetic materials   铁磁性材料

Magnetic permeability is very much greater than metals that have a high attraction to permanent magnets; e. g. iron, steel, cobalt, nickel. MPI is carried out only on ferromagnetic materials with large values of μ.

2. paramagnetic materials   顺磁性材料

Magnetic permeability μ is slightly greater than metals that have only a very slight attraction to magnetism, usually insufficient to attract a permanent magnet, e. g. aluminum, titanium.

3. diamagnetic materials   逆磁性材料

Magnetic permeability is less than metals that repel magnetic fields, e. g. , copper, brass, bismuth.

4. circular magnetization   周向磁化

The electric current is passed through the workpiece to produce a circular magnetic field.

5. central conductor magnetization　中心导体磁化法

The inside surface of a tubular workpiece can be examined by a magnetic field produced by a conductor threaded through the workpiece.

6. coil magnetization　线圈磁化法

Workpiece placed in a current-carrying coil becomes magnetized along the coil axis.

7. yoke magnetization　磁轭磁化

A magnetizing field is produced by applying the ends of a U-shaped soft-iron armature to the workpiece.

8. magnetic lines of force　磁力线

This is used to describe a magnetic field. A magnetic line of force is such that its direction at every point is the same as the direction of the force that would act on a small magnetic pole placed at the point. A magnetic line of force is defined as starting from a north pole and ending on a south pole. Magnetic lines of force are always perpendicular to the direction of electric current flow.

9. magnetic induction or magnetic flux density (B)　磁感应强度或磁通密度 (B)

The strength of the magnetic field in the part: in units of Tesla (SI) or Gauss (CGS emu). The ratio $B/H = \mu$ is the magnetic permeability of the material; for ferromagnetic materials such as iron μ is a function of H, and is a maximum where B increases most rapidly with H, that is at the knee of the magnetization hysteresis curve.

10. magnetic leakage field (flux leakage field)　漏磁场

The magnetic field that leaves or enters the surface of a part at a discontinuity.

11. magnetizing force (H)　磁场强度

The strength of the magnetic field in air (or vacuum): in units of amps/meter (SI) or oersted (cgs emu).

12. coercive force　矫顽力

The magnetic force H required to bring the magnetic flux density B of a material to zero; the magnetic field required to demagnetize a solid.

13. residual field inspection　剩磁检验

Application of magnetic particles to the part after the magnetization current has been removed.

14. residual magnetic field (retentivity, remanence)　剩磁场

The field that remains in a magnetized material after the magnetizing force has been removed.

15. skin effect   集肤效应

The tendency of alternating currents to flow near to the surface of a material.

16. hysteresis   磁滞

The lag of the magnetization of a ferromagnetic behind the applied magnetic field.

学习笔记

Chapter 12　Penetrant Testing

Liquid penetrant testing is a nondestructive method for finding discontinuities that are open to the surface of solid and essentially nonporous materials. Indications of flaws can be found regardless of the size, configuration, internal structure or chemical composition of the workpiece being inspected and regardless of flaw orientation. Liquid penetrants can seep into (and be draw into) various types of minute surface openings (reportedly, as fine as 0. 1 μm, or 4 μin width) by capillary action. Because of this, the process is well suited for detection of all types of surface cracks, laps, porosity, shrinkage areas, laminations and similar discontinuities. It is used extensively for inspection of wrought and cast products of ferrous and nonferrous metals, powder metallurgy parts, ceramics, plastics, and glass objects.

Regardless of the type penetrant used, and regardless of other variations in the basic process, liquid-penetrant testing requires at least five essential steps:

(1) **Surface Preparation**　All surfaces of a workpiece must be thoroughly cleaned and completely dried before it is subjected to liquid penetrant testing.

(2) **Penetration**　After the workpiece has been cleaned, liquid penetrant is applied in a suitable manner so as to form a film of the penetrant over the surface for at least 13 mm beyond the area being inspected.

(3) **Removal of Excess Penetrant**　Next, excess penetrant should be removed from the surface. Uniform removal of excess penetrant is necessary for effective inspection, but overcleaning must be avoided.

(4) **Development**　A developing agent is applied so that it forms a film of over the surface. The developer also provides a uniform background to assist visual inspection.

(5) **Inspection**　After being sufficiently developed, the surface is visually examined for indications of penetrant bleedback from surface openings.

These five essential operations are shown schematically for the water-washable system in the following figure. The operations are similar for the other liquid-penetrant systems.

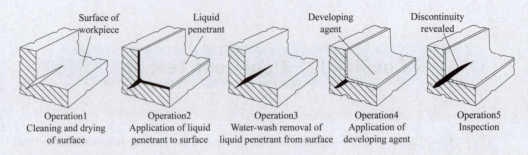

Five essential operations for liquid-penetrant testing using the water-washable system

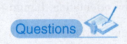

1. What is the principle of liquid penetrant testing?
2. What are the basic steps of liquid penetration testing?

Indications of flaws can be found regardless of the size, configuration, internal structure or chemical composition of the workpiece being inspected and regardless of flaw orientation.

无论缺陷的尺寸、形状、被检工件内部组织结构或化学成分以及缺陷取向如何，其缺陷的痕迹都能被发现。

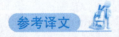

渗透检测

液体渗透检测是一种主要用于发现固体和基本无孔材料表面的不连续、开口类缺陷的无损检测方法。无论缺陷的尺寸、形状、被检工件内部组织结构或化学成分以及缺陷取向如何，其缺陷的痕迹都能被发现。液体渗透剂可以通过毛细作用渗入（并被吸入）各种类型的微小表面开口类缺陷中（据报道，细小至 $0.1~\mu m$ 或 $4~\mu in$ 宽）。正因为如此，这种方法非常适合检测所有类型的表面裂纹、搭接、孔隙、缩孔、层压和类似的不连续性缺陷。渗透检测广泛应用于检验铁质和非铁金属的锻件、铸件产品、粉末冶金零件、陶瓷、塑料和玻璃制品。

不管使用何种类型的渗透剂和基本工艺的其他变化，液体渗透检测至少需要五个基本步骤：

（1）**表面处理**　在进行液体渗透检测之前，工件的所有表面必须彻底清洁并完全干燥。

（2）**渗透**　清洁工件后，以适当的方式施加液体渗透剂，以便在被检区域及其以外至少 13 mm 的表面上覆盖有渗透剂。

（3）**去除多余渗透剂**　应清除表面多余的渗透剂。均匀去除多余的渗透剂是有效检测的必要条件，但是必须避免过洗。

（4）**显像**　施加显像剂使其在工件表面上形成一层膜。显像剂还提供了一个统一的背景，以帮助目视检测。

（5）**检测**　充分显影后，目视检查表面是否有渗透剂从表面开口回流的迹象。

下图所示为适用于水洗系统的这五个基本操作。其他液体渗透剂检测系统的操作与其类似。

Reading Material

Basic Knowledge of Liquid Penetrant Testing

Capillarity, or capillary attraction is the action by which the surface of a liquid, where it is in contact with a solid, is elevated or depressed.

The materials, processes, and procedures used in liquid penetrant testing are designed to facilitate capillarity and to make the results of such action visible and capable of interpretation.

Surface Wetting and Contact Angle　Liquid penetrant tests for detection of surface discontinuities depend on the ability of the penetrant to enter the discontinuities and remain there as a penetrant entrapment.

Capillarity　Nonetheless, penetration depth is still proportional to penetrant surface tension, and inversely proportional to crack width and penetrant density.

Visible Dye Penetrant Kit　The visible dye penetrant test kit is light in mass and contains all the materials necessary for test.

Stationary Equipment　The stationary equipment used in liquid penetrant testing ranges from the simple to fully automatic systems and varies in size, layout, and arrangement depending on the requirements of specific tests.

Types of Penetrants   Fluorescent penetrants contain a dye or several dyes that fluoresce when exposed to ultraviolet radiation. Visible penetrants contain a red dye that provides high contrast against the white developer background.

Characteristics of Penetrant Materials   A penetrant must spread easily over the surface of the material being inspected to provide complete and even coverage.

Testing methods Classification   Method A—water washable；Method B—post emulsifiable， lipophilic；Method C—solvent removable；Method D—post emulsifiable， hydrophilic.

Basic Processing Steps   Surface preparation—penetrant application—penetrant dwell—excess penetrant removal—developer application—indication development—inspection—clean surface.

Excess Penetrant Removal   This is the most delicate workpiece of the inspection procedure because the excess penetrant must be removed from the surface of the workpiece while removing as little penetrant as possible from defects.

Indication Development   The developer is allowed to stand on the part surface for a period of time sufficient to permit the extraction of the trapped penetrant out of any surface flaws.

Water Washable Liquid Penetrant Process   To ensure that testing is reliable and reproducible and remains sufficiently sensitive for the purpose intended， the water washable liquid penetrant test procedure includes the following operational steps：

Surface preparation—penetrant application—penetrant dwell—excess penetrant removal—developer application—indication development—inspection—clean surface.

Lipophilic Postemulsification Liquid Penetrant Process   Because the liquid penetrant does not contain an emulsifier， an additional step of applying emulsifier and carefully controlling emulsifier dwell time must be provided to make the excess surface liquid penetrant removable by water spray washing.

参考译文

渗透检测的基础知识

毛细作用，或毛细吸附是作用在与固体接触的液体表面上的力，这种力使液体表面上升或下降。

渗透检测的材料、工艺和程序的设计都是为了促进毛细现象，并使这种毛细现象的结果可视、便于解释。

表面润湿和接触角　渗透检测能否检测到表面的不连续性，取决于渗透剂进入不连续性并在不连续性中保留的能力。

毛细现象　然而，渗透深度正比于渗透剂的表面张力，反比于裂纹的宽度以及渗透液的浓度。

着色检测器材　着色检测器材质量轻，包括检测所需的所有材料。

固定式设备　渗透检测应用的固定式检测系统从简单型到全自动控制型不等，根据具体的检测要求，固定式设备要在尺寸、布局和安排上进行调整变化。

渗透剂的种类　荧光渗透剂包含一种或多种染料，当暴露在紫外线辐射下会发出荧光。着色渗透剂包含红色染料，在白色的显像剂背景下可以得到很好的对比度。

渗透材料的特性　渗透剂必须很容易施加到被检工件表面，并能完整而均匀地覆盖。

渗透检测方法分类　方法 A：水洗型；方法 B：亲油性后乳化型；方法 C：溶剂去除型；方法 D：亲水性后乳化型。

渗透检测基本检测步骤　表面准备——施加渗透剂——渗透剂保留——去除多余的渗透剂——施加显像剂——显示显像——检验——表面清洗。

去除多余的渗透剂　这是渗透检测过程中最慎重的一个步骤，需要将工件表面多余的渗透剂去除掉，而尽可能少地将缺陷中的渗透剂去除。

显示显像　显像剂应在工件表面停留足够的时间以保证表面缺陷中的渗透剂被吸附出来。

水洗型渗透检测工艺　为了确保检测的可靠性、可重复性以及达到所要求的检测灵敏度，水洗型渗透检测的程序包括以下检测步骤。

表面准备——施加渗透剂——渗透剂保留——去除多余的渗透剂——施加显像剂——显示显像——检验——表面清洗。

亲油性后乳化渗透检测工艺　由于渗透剂中不含乳化剂，所以，需要增加一个额外的施加乳化剂的步骤，乳化剂的停留时间必须小心控制，以便喷水将表面多余的渗透剂去除。

New Words

1. nonporus   *adj.* 无（细，气）

2. configuration   *n.* 形状

3. orientation   *n.* 方向

4. lamination   *n.* 层压结构

5. ceramics   *n.* 陶瓷

6. overcleaning   *n.* 过洗

7. developer   *n.* 显像剂，显色剂

Phrases and Expressions

1. ferrous and nonferrous metal   铁质和非铁金属

2. essential step   基本步骤

Glossary of Terms for Penetrant Testing

1. wet（wetting ability）   润湿

The capacity of a fluid to coat or cover a material in a thin layer.

2. capillary action   毛细管作用

The movement of liquid within narrow spaces that is caused by surface tension between the liquid and the substrates materials that adversely affects the performance of penetrant materials.

3. penetrant   渗透剂

A solution or suspension of dye（visible or fluorescent）that is used for the detection and evaluation of surface-breaking discontinuities.

4. penetrant   渗透剂

A penetrant is characterized by an intense color, usually red.

5. dwell time   停留时间，渗透时间

The total time that the penetrant is in contact with the test surface, including the time required for application and the drain time.

6. drying time　干燥时间

The time required for a cleaned, rinsed or wet developed part to dry.

7. emulsifier　乳化剂

A liquid that interacts with an oily substance to make it water-removable.

8. hydrophilic emulsifier　亲水性乳化剂

A water-based liquid used in penetrant examination that interact with the penetrant oil, rendering it water-removable.

9. lipophilic emulsifier　亲油性的乳化剂

An oil-based liquid used in penetrant testing that interacts with the penetrant oil rendering it water-removable.

10. emulsification time　乳化时间

The time that an emulsifier is permitted to remain on the part to combine with the surface penetrant prior to removal.

11. developer　显像剂

A material that is applied to the test surface to accelerate bleedout and to enhance the contrast of indications.

12. development time　显像时间

The elapsed time between the application of the developer and the examination of the workpiece.

13. surface tension　表面张力

The force acting tangentially to the surface to reduce the surface area to a minimum. It acts to draw a volume of liquid into a sphere.

14. precleaning　预清洗

The removal of surface contaminants from the test part so that they will not interfere with the examination process.

15. water-removable penetrant　水洗型渗透剂

A penetrant with a built-in emulsifier.

16. postemulsifiable penetrant　后乳化渗透剂

A penetrant that requires the application of a separate emulsifier to render the excess surface penetrant water-washable.

17. solvent removable penetrant　溶剂去除型渗透剂

A penetrant so formulated that most of the excess surface penetrant can be removed by wiping, with the remaining surface penetrant traces removable by further wiping with a cloth or similar material lightly moistened with a solvent remover.

18. penetrant comparator block　渗透对比试块

An intentionally flawed workpiece having separate but adjacent areas for the application of different penetrant materials so that a direct comparison of their relative effectiveness can be obtained.

19. solvent remover　溶剂去除剂

A volatile liquid used to remove excess penetrant from the surface being examined.

20. postemulsification　后乳化

The application of a separate emulsifier after the penetrant dwell time.

21. postcleaning　后清洗

The removal of residual penetrant test materials from the workpiece after the penetrant test has been completed.

22. fluorescence　荧光，荧光性

The property of materials to absorb a high-energy, short-wave length light such as ultraviolet light, and then remit the absorbed energy as a longer-wavelength, lower energy light (visible light).

23. ultraviolet lamp　紫外灯

A lamp that produces principally ultraviolet light that cannot be seen by the human eye. Some visible light often escapes the filter on the lamp and can be seen.

Chapter 13   Eddy-Current Testing

When electrically conductive material is subjected to an alternating magnetic field, small circulating electric currents are generated in the material (Fig. 13−1). These so-called eddy currents are affected by variations in conductivity, magnetic permeability, mass, and homogeneity of the host material. Conditions that affect these characteristics can be sensed by measuring the eddy-current response of the workpiece.

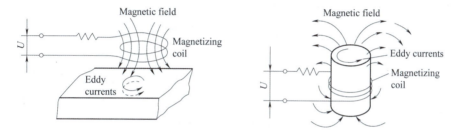

Fig. 13−1   Eddy-current testing

The eddy currents induced into the workpiece interact with the magnetic field of the exciting coil, thereby influencing the impedance, which is the total opposition to the flow of current from the combined effect of resistance, inductance, and capacitance of the coil. [1]

By measuring the impedance of the exciting coil, or in some cases a separate indicating coil, eddy-current testing can detect cracks, voids, inclusions, seams, and laps. The best results are obtained when the current flow is at right angles to the flaw (Fig. 13−2).

According to a NASA report, eddy-current testing is not as sensitive to small, open flaws as is liquid penetrant. However, it does not require cleanup operations and is generally faster. Compared with magnetic particle testing, eddy-current testing is not as sensitive to small flaws but they work equally as well on ferromagnetic and nonferromagnetic material.

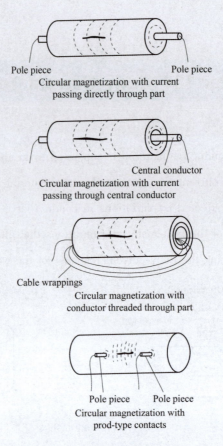

Pole piece   Pole piece
Circular magnetization with current
passing directly through part

Central conductor
Circular magnetization with current
passing through central conductor

Cable wrappings
Circular magnetization with
conductor threaded through part

Pole piece   Pole piece
Circular magnetization with
prod-type contacts

Fig. 13-2   Magnetization of the workpiece may vary depending upon
the application, but for optimum indications the direction of the magnetic fields
should be nearly right angles to the fault[2]

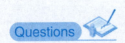

Questions

1. What is the principle of eddy-current testing?

2. Describe the uses of eddy current testing.

3. How does eddy-current testing differ from magnetic particle testing?

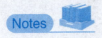

Notes

①The eddy currents induced into the workpiece interact with the magnetic field of the exciting coil, thereby influencing the impedance, which is the total opposition to the flow of current from the combined effect of resistance, inductance, and capacitance of the coil.

工件中感应的涡流与励磁线圈中的磁场相互作用从而影响涡流阻抗，这种阻抗对涡流流向的影响与线圈中的电阻、电感、电容共同对电流的影响是完全相反的。

在本句中，which 引导非限定性定语从句。

②Magnetization of the workpiece may vary depending upon the application，but for optimum indications the direction of the magnetic fields should be nearly right angles to the fault.

尽管工件的磁化随着它的用途的不同而不同，但是要得到最佳显示效果，其磁场方向几乎与缺陷成直角。

句中 nearly 作副词，可译为"几乎，差不多"。

参考译文

涡流检测

当导电材料受到一个交变磁场的作用时，就会在材料中产生一个小的环路电流（图13-1）。这就是所谓的涡流因导电率、磁通量、质量和基体材料的均匀性等特性的不同而不同。影响这些特性的条件可以通过测量工件感应的涡流感知。

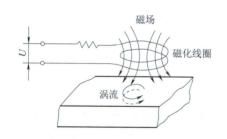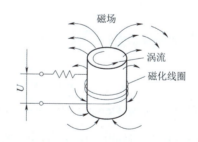

图 13-1 涡流检测

工件中的感应涡流与励磁线圈中磁场相互作用从而影响涡流阻抗，这种阻抗对涡流流向的影响与线圈电阻、电感、电容共同对电流的综合影响是完全相反的。

通过测量励磁线圈的阻抗或有时候表示独立线圈的阻抗，涡流检测能检测到裂纹、孔洞、夹杂、焊缝和搭接焊缝等。当电流与缺陷成直角时，就可得到最好的结果（图13-2）。

根据美国航空航天局的一个报告，涡流检测对小的开放的缺陷不如液体浸

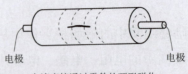

电极　　　　　　　　　　　电极

电流直接通过零件的环形磁化

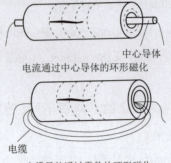

中心导体

电流通过中心导体的环形磁化

电缆

电缆导体通过零件的环形磁化

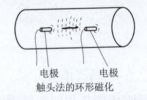

电极　　　　电极

触头法的环形磁化

图 13-2　工件的磁化可能取决于应用，但为了获得最佳的显示，磁场的方向应该与缺陷方向垂直

入灵敏。不管怎样，它是不需要清洗设备的，并可快速进行检测。与磁粉探伤相比，涡流检测对小缺陷不敏感，但对铁磁和非铁磁材料同样有效。

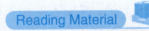

Eddy Current Testing Method

Measures or Defects　Surface and **subsurface** cracks and **seams**, alloy content, heat treatment variations, wall thickness, coating thickness, cracks depth, metal sorting.

Applications　Tubing, wire, ball bearings, spot checks on all types of surfaces, proximity gauge, metal detector.

Advantages　No special operator skills required, high speed, low cost, automation possible for symmetrical parts, permanent record capability for symmetrical parts, no couplant or probe contact required.

Limitations　Conductive materials, shallow depth of penetration (thin walls

only), masked or false indications caused by sensitivity to variations, such as part geometry, reference standards required, permeability.

Basic Knowledge of Eddy-Current Testing

If another electrical conductor (test material) is brought into the close proximity to this changing magnetic field, eddy current that flow in a circular path will be induced in this second conductor due to mutual induction.

By measuring changes in the resistance and inductive reactance of the coil, information can be gathered about the test material. This information includes the electrical conductivity and magnetic permeability of the material, and whether it contains cracks or other defects.

Skin Effect and Standard Penetration Depth　The depth that eddy currents penetrate into a material is affected by the frequency of the excitation current and the electrical conductivity and magnetic permeability of the workpiece. The depth at which eddy current density has decreased to $1/e$, or about 37% of the surface density, is called the standard depth of penetration.

Lift-off Effect　Lift off can be used to make measurements of the thickness of nonconductive coating such as paint that hold the probe a certain distance from the surface of the conductive material.

Selection of Inspection Frequencies　At the lower frequencies, the penetration δ is relatively high, but the sensitivity to the detection of discontinuities is relatively low; the reverse is true at higher frequencies.

In the case of a cylindrical shape of radius r, frequency selection depends on the characteristic frequency (also known as limit frequency) which is derived by considering the strength of the magnetic field penetrating into the workpiece.

End Effects　Eddy currents are distorted at the ends of the workpiece. Inspections should not be within the coil diameter from an edge depending on factors such as test coil size and frequency.

Instrument　Most eddy current instruments are dedicated to a particular application, such as the detection of cracks, inspection of tubes, metal sorting, or the determination of coating thickness or conductivity.

Probe　Surface probes are usually designed to be held with hands and intended to be used in contact with the surface of the workpiece.

Reference Standard   In eddy current testing, the use of reference standards in setting up the equipment is particularly important since signals are affected by many different variables and slight changes in equipment setup can drastically alter the appearance of a signal.

Multifrequency   One method of overcoming interfering signals from other conductors in the vicinity of the workpiece under test is by the use of multi-excitation frequencies.

参考译文

涡流检测的基础知识

如果另一个导体（被检材料）靠近变化的磁场，由于互感就会在该导体中感应出沿圆形流动的涡流。

通过测量线圈中阻抗和感抗的变化，就可以得到被测材料中的信息。这些信息包括被测材料的电导率、磁导率，是否存在裂纹以及其他缺陷。

集肤效应和标准透入深度   涡流在材料中的渗透深度受激励电流的频率以及材料的电导率和磁导率的影响。电流密度下降到表面的 1/e 或大约 37% 的深度称为标准透入深度。

提离效应   提离效应可以用于测量像漆层这样的非导电涂层的厚度，因为涂层使探头与导体表面之间保持一定的距离。

检测频率选择   频率较低时，渗透深度比较深，但是检测灵敏度相对要低，反之亦然。

对于半径为 r 的圆柱体，检测频率的选择取决于特征频率（也称极限频率），特征频率是通过进入工件中的磁场强度推导出来的。

边缘效应   涡流在工件边缘会发生畸变。涡流的检测范围取决于检测线圈的尺寸和频率。

检测仪器   大部分涡流检测仪器都专注于特定的应用，如检测裂纹、检测管材、金属分选或者测量涂层厚度或电导率。

探头   表面探头通常设计成手持式，目的是方便与工件表面接触。

参考试块   对于涡流检测，在检测设备的安装调试过程中参考试块的使用非常重要，这是因为信号会受到很多因素的影响，设备安装调试过程中的微小的变化都会对信号产生很大的影响。

多频检测技术　克服邻近被检工件的其他导体引起的干扰信号的一个方法就是使用多频检测法。

New Words

1. conductivity　*n.* 传导率，导电率
2. permeability　*n.* 磁导率
3. mass　*n.* 物质
4. homogeneity　*n.* 同种，均匀性，均质性
5. host　*n.* 基质，晶核
6. exciting　*adj.* 激励的，励磁的
7. coil　*n.* 线圈，感应圈
8. impedance　*n.* 阻抗
9. inductance　*n.* 电感系数
10. capacitance　*n.* 电容
11. optimum　*adj.* 最佳的，最有利的
12. void　*n.* 空隙，空洞，白点
13. seam　*n.* 焊缝
14. lap　*n.* 搭接缝

Phrases and Expressions

1. electrically conductive material　导电材料
2. be subjected to　使……受到，在……条件下
3. magnetic field　磁场
4. eddy current　涡流
5. exciting coil　励磁线圈
6. magnetic particle testing　磁粉检测
7. ferromagnetic material　铁磁材料
8. nonferromagnetic material　非铁磁材料

Glossary of Terms for Eddy Current

1. eddy current 涡流

Closed loop alternating current flow induced in a conductor by a time-varying magnetic field.

2. electromagnetic coupling 电磁耦合

Electromagnetic interaction between two or more circuits. In eddy current examination the workpiece to be tested becomes a circuit. See also magnetic coupling; mutual inductance.

3. encircling coils 外穿式线圈

Coils that surround the workpiece to be tested; also known as annular, circumferential, or feed-through coils.

4. ID coil 内穿式线圈

A coil assembly for internal testing by insertion into the test piece; also known as inside coils, inserted coils, or bobbin coils.

5. bobbin probe or coil 线轴探头或线圈

Eddy current probe designed to inspect the interior of a tube; also called an interior or internal probe, as it is inserted into the tube.

6. surface probe 放置式线圈（探头）

Eddy current probe designed for testing surfaces; the coil is typically wound in a pancake shape.

7. reference coil 参考线圈

Coil that enables bridge balancing in absolute probes; its impedance is close to the test coil impedance, but does not couple to the test material.

8. absolute probe 绝对式探头

Probe having a single sensing coil; it may or may not have a separate drive coil.

9. differential probe 差动式探头

Probe having two sensing coils located side-by-side. The probe may or may not have a separate drive coil.

10. conductivity 导电性

Measure of the ability of a material to conduct electrical current.

11. resistance 电阻

Ability of a material to resist the flow of an electric current; measured in ohms.

12. impedance 阻抗

The opposition of a circuit to the flow of an alternating current: the complex quotient of voltage divided by current.

13. capacitive reactance 容抗

Reactance in a circuit that opposes any changes in voltage.

14. lift-off effect 提离效应

The effect of a change in magnetic coupling between the workpiece and the test coil as the distance between them is varied.

15. edge effect 边缘效应

Signal obtained when a surface probe approaches a sample's edge.

16. skin effect 趋肤效应

Phenomenon where currents and magnetic fields are restricted to the surface of a test sample: increasing test frequency reduces penetration (see skin depth).

Chapter 14   Other NDT Methods

Acoustic Emission   The acoustic emission (AE) testing method is a unique nondestructive testing method where the material being tested generates signals that warn of impending failure. Acoustic emission testing is based on the fact that solid materials emit sonic or ultrasonic acoustic emissions when they are mechanically or thermally stressed to the point where deformation or fracturing occurs.[1] During plastic deformation, dislocations move through the material's crystal lattice structure producing low-amplitude AE signals, which can be measured only over short distances under laboratory conditions. The AE test method detects, locates, identifies, and displays flaws data for the stressed object the moment the flaw is created. Therefore, flaws can not be retested by the AE method. In contrast, ultrasonic testing detects and characterizes flaws after they have been created. Almost all materials produce acoustic emissions when they stressed beyond their normal design ranges to final failure.

Time of Flight Diffraction   The time of flight diffraction (TOFD) ultrasonic testing method is relatively new and was first developed at Harwell laboratory in the late 1970 s by Maurice Silk. TOFD testing has been gaining in profile over the last three or four years with much interest focussing on whether or not it can be used to replace more established NDT methods. Recent survey shows that the annual average growth rate (AAGR) of the TOFD market is 10%–20% higher than other NDT techniques. The TOFD method is gaining and increasing popularity because of its high probability of detection, low false call rate, portability and most important intrinsic accuracy in flaw sizing, especially in depth. There is another NDT method called radiography testing (RT/X-ray) usually employed for flaw sizing. It should be noted that RT/X-ray technique shows better accuracy for lateral flaw sizing but it demonstrates insufficient accuracy in depth assessment. As the standards for radiation safety become tighter by the new European law, many NDT companies are trying to substitute X-ray inspection by TOFD method for cost effectiveness and mostly for safety and environment protection reasons.

The most significant distinction between TOFD and the other UT methods is that it monitors only forward-scattered diffracted energies from the tips of defects rather than reflected ultrasonic energies. Two wide beam angle probes are used in transmitter-receiver mode. Broad beam probes are used so that the entire crack area is flooded with ultrasound and, consequently, the entire volume is inspected using a single scan pass along the inspection line.[2] Because the method relies on detection of the forward scattered diffracted signals originating at the flaw edges, precise measurement of flaw size, location, and orientation is possible.

Thermal Nondestructive Testing Thermal nondestructive testing is a general term for the various methods used to detect flaws and an undesirable distribution of heat during service. The methods used normally can be classified as either noncontact or contact. Noncontact methods depend on thermally generated electromagnetic energy radiated from the part being tested. At moderate temperatures this is predominantly the infrared region. Because of this, infrared testing is the most important branch of noncontact thermal testing. Direct contact methods place a thermally sensitive device or material in physical and thermal contact with the test part.

The noncontact and contact methods can be further classified as either thermographic or thermometric methods. Thermographic methods depend on the thermal gradients that occur on the surface of the part. When these methods are used, a map of the equal temperature contours is obtained. With thermometric methods, a precise value of the temperature is obtained.

Leak Testing Leak testing is a form of nondestructive testing capable of determining the existence of leak sites and, under proper conditions, measuring the quantity of material passing through these sites. The term leak refers to a hole or passage through which a fluid passes in either a pressurized or e-vacuated system. Two types of leaks exist: real leaks and virtual leaks. A real leak is a discrete hole or passage through which a fluid may flow. Virtual leaks are sources of gradual desorption of gates from surfaces or components within a vacuum system.

Industrial Computed Tomography Industrial computed tomography (ICT) is a unique nondestructive evaluation technique for inspecting interior structures of test objects. While computed tomography has long been used by radiologists for medical diagnostics applications, industrial CT brings the same technique to inspection

applications. By utilizing the penetrating capabilities of X-rays, the technology generates cross-sectional images of an object for detailed analysis. The cross sectional images and the 3D views (volume rendered by processing of cross-sectional images) provide the inspection specialist with a wide range of options for detecting cracks and air voids; measuring distances, areas and volumes; and locating the problem areas in hard-to-access components. All analysis is done using a workstation computer integrated within the CT imaging system that runs viewer software to visualize the tomography data.

Visual and Optical Inspection　　Visual inspection continues to develop as an American Society for Nondestructive Testing (ASNT) evaluation method. In the past, visual inspection was considered highly subjective in nature and provided little or no hardcopy documentation. Successful results depended on trained operators, cleanliness and condition of the test object, quality of the optical instrument, and proper illumination of the test part. These factors are still important today, but equipment has become much more sophisticated.

Visual examination is one of the most basic nondestructive evaluation methods. Quality control inspectors follow procedures that range from simply looking at a part to see surface imperfections to performing various gauging operations, which assure compliance with acceptable physical standards. Today's optical systems may include special probes, spectrometers, and realtime imaging and analysis using notebook computers.

High-tech pipe weld inspection systems often combine multiple NDT methods. For example, remote magnetic pipe crawlers can be easily equipped with compact video cameras and eddy current or ultrasonic probes to visually inspect the weld and determine the extent of defects.

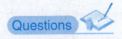

1. What are other nondestructive testing methods?

2. What are the advantages of TOFD?

3. What is the advancement of industrial computed tomography?

Notes

①Acoustic emission testing is based on the fact that solid materials emit sonic or ultrasonic acoustic emissions when they are mechanically or thermally stressed to the point where deformation or fracturing occurs.

声发射检测方法基于一个事实：固体材料在受到机械或热应力后，当到达发生变形或破裂的临界点时，将会发射出声波或超声波。

②Two wide beam angle probes are used in transmitter-receiver mode. Broad beam probes are used so that the entire crack area is flooded with ultrasound and, consequently, the entire volume is inspected using a single scan pass along the inspection line.

使用两个工作于发射-接收模式的宽光束探头，使得整个裂纹区域充满超声波，并且使用单次扫描沿着检查线通道检查整个体积。

参考译文

其他无损检测方法

声发射　声发射（AE）检测方法是一种通过检测材料发出的故障预警信号来进行无损检测的独特方法。声发射检测方法基于一个事实：固体材料在受到机械或热应力后，当到达发生变形或破裂的临界点时，将会发射出声波或超声波。在塑性变形过程中穿过材料的晶格结构的位错会产生低振幅的声发射信号，此信号只能在实验室条件下的短距离处测量。声发射检测在承压物体出现缺陷的瞬间对缺陷数据进行探测、定位、识别和显示。因此，不能使用声发射检测方法对缺陷进行复检。与此相反，超声检测方法是在缺陷产生后对其检测和鉴定。几乎所有材料所承受的压力在超出其正常设计范围到完全破坏时，都会产生声发射。

衍射时差法　衍射时差法（TOFD）相对较新，最早于1977年末由Maurice Silk在Harwell实验室开发。衍射时差法在过去的三四年里有着越来越多的研究，研究者们更加关注该技术是否可以用来取代更成熟的无损检测方法。最近的调查显示，衍射时差法的市场的年平均增长率比其他NDT技术高10%~20%。衍射时差法由于其具有高检测概率、低误报率、便携，以及在缺陷尺寸（尤其是深度）的内在精度这一个最重要的优势而越来越受欢迎。射线检测（RT／X射线）是另一种通常用于缺陷尺寸测量的无损检测方法。应该注意的是，射线检测具有更好的横向缺陷尺寸精度，但它在深度评估方面的准确性不足。随着欧洲新法律对辐射安全的标准越来越严格，许多无损检测公司出于安全和环境保护的原因，正在尝试用衍射时差法替代射线检测以获得成本效益。

衍射时差法和其他超声检测方法之间最显著的区别是，它仅监测来自缺陷尖端的前向散射衍射能量，而不是检测反射超声能量。使用两个工作于发射–接收模式的宽光束探头，使得整个裂纹区域充满超声波，并且使用单次扫描沿着检查线通道检查整个体积。因为该方法依赖于源于缺陷边缘的前向散射衍射信号的检测，所以可以实现缺陷尺寸、位置和方向的精确测量。

热成像无损检测　热成像无损检测是现存检测不期望热量分布及缺陷的各种方法的总称。热成像通常使用的方法可以分为非接触式方法或接触式方法。非接触式方法取决于从被测零件辐射的热产生的电磁能量。在中等温度下，电磁辐射主要位于红外区域。因此，红外检测是非接触式热检测中最重要的分支。接触式方法是使热敏器件或材料与测试零件进行物理和热接触。

非接触式方法和接触式方法可以进一步分为热成像方法与测温方法。热成像方法取决于零件表面的热梯度，使用热成像方法可以得到被测零件的等温线；使用测温方法可以获得精确的温度值。

泄漏检测　泄漏检测是一种能够确定泄漏点的无损检测方法，并在适当的条件下通过这些点测量材料的质量。术语"泄漏"是指在加压或真空系统中流体通过的孔或通道。存在两种类型的泄漏：真泄漏和虚泄漏。真泄漏是流体可以流过的离散孔或通道；虚泄漏是从真空系统内的表面或部件逐渐退吸的泄漏源。

工业计算机断层扫描　工业计算机断层扫描（ICT）是一种用于检测物体内部结构的独特的无损检测技术。计算机断层扫描长期以来一直被放射科医师用于医疗诊断应用，工业CT也将相同的技术应用在检测中。通过利用X射线的穿透能力，工业CT生成物体的横断面图像用于进行详细分析。横断面图像和三维视图（通过横断面图像处理得到的三维体）为检测专家提供了多种选择：检测裂纹和气隙，测量距离、面积和体积，在难以访问的组件中对问题区域进行定位。所有分析均使用集成在CT成像系统内的工作站计算机上完成，该计算机系统可以运行观察器的软件从而使断层摄影数据可视化。

目视和光学检测　目视检查作为美国无损检测协会（ASNT）的评估方法得到

了持续的发展。在过去，目视检查在本质上被认为是一种高度主观性的技术，并且只能提供很少或不能提供有效的文档记录。成功的目视检查结果取决于操作员的训练程度、测试对象的清洁度和状况、光学仪器的质量以及测试零件的适当照明。这些因素在今天仍然很重要，不过现在的检测设备变得更加复杂。

目视检查是最基本的无损评估方法之一。质量控制检验员按照程序从简单地观察零件表面的不连续性到执行各种测量操作，以确保零件符合实际的接受准则。今天的光学系统可能包括特殊探头、光谱仪，以及使用笔记本电脑的实时成像和分析程序。

高科技管道焊缝检测系统通常汇集了多种无损检测方法。例如，可以在远程磁力管道爬行器上配备紧凑型摄像机和涡流或超声波探头，从而可用目视检查方法检查焊缝并确定缺陷的程度。

 Reading Material

Introduction to Other NDT Methods

Visual Testing   Visual testing（VT）is perhaps the oldest and most widely used inspection technique. Often, the eyes of the inspector are the only "equipment" used for the inspection.

Environmental factors that can affect a visual inspection include atmosphere, cleanliness of the object being inspected, and the position of object in relation to the inspector.

Infrared and Thermal Testing   Because temperature is, by far, the most measured and recorded parameter in industry, it is no surprise that applications for temperature measurement and thermography are found in virtually every aspect of every industry.

Leak Testing   It is a very useful technique for locating leaks, and then repairing, to reduce or eliminate leakage in a system. Leakage monitoring is a long term, continuous test ensuring that containment of a vessel has not been breached.

Acoustic Emission Testing   Acoustic emission testing is a nondestructive method with demonstrated capabilities for monitoring structural integrity for detecting leaks and incipient failures in mechanical equipment and for characterizing material behavior.

Acoustic emission testing has been demonstrated in various applications on aircraft, including crack growth detection in flight and global monitoring of entire airframes for crack growth.

Time of Flight Diffraction   The defect sizing method based on the measurements of time difference between the diffracted signals from the crack tips (edges) is called time of flight diffraction method.

Industrial Computed Tomography（**ICT**）   Computed tomography differs from conventional radiographic imaging in that it uses X-ray transmission information from numerous angles about an object to computer reconstruct cross sectional images (that is, slices) of the interior structure.

Computed Radiography   Computed radiography (CR) uses very similar equipment to conventional radiography except that in place of a film to create the image, an imaging plate (IP) made of photostimulable phosphor is used.

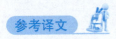

其他无损检测方法的介绍

目视检测   目视检测或许是历史上最古老、应用最广泛的检测技术。通常，检测人员的眼睛就是检测过程中唯一使用的设备。

影响目视检测的环境因素包括大气、被测工件的清洁度以及工件与检测人员的相对位置。

红外热成像检测   因为到目前为止，温度是工业中测量与记录最广泛的参数，所以温度测量与热成像技术在几乎每个工业领域都得到应用就不足为奇了。

泄漏检测   这项技术对确定泄漏位置非常有用，在定位的基础上再进行修理，从而减少或消除系统的泄漏。泄漏的监控是一个长期、持续的检测任务，需要确保容器的密封性，避免容器发生破裂。

声发射检测   声发射检测是一种无损检测方法，已经证明该项技术具有监控结构完整性、检测机械设备的泄漏和初期损伤以及表征材料性能的能力。

声发射检测已经被证明在飞行器检测中有多种应用，包括飞行中裂纹扩展检测，整体机身裂纹扩展的全面监控。

衍射时差法超声检测技术   衍射时差法超声检测技术是一种通过测量裂纹尖端衍射信号的时间差来确定缺陷尺寸的超声检测方法。

工业计算机断层扫描   不同于常规的 X 射线照像成相技术，计算机层析技术运用穿过工件多角度的 X 射线的信息，通过计算机重建工件内部横断面图像（即切片）。

计算机 X 射线照相术　CR 所用的设备与常规 X 射线照相术差不多，不同之处在于使用光激励荧光制作的成像板取代了胶片成像。

Phrases and Expressions

1. acoustic emission（AE）　声发射

2. time of flight diffraction（TOFD）　衍射时差法

3. crack tip　裂纹尖端

4. thermal nondestructive testing　热成像无损检测

5. infrared testing　红外检测

6. leak testing　泄漏检测

7. industrial computed tomography（ICT）　工业计算机断层扫描成像，工业计算机层析技术

8. visual inspection　目视检查

9. notebook computer　笔记本电脑

Chapter 15　Destructive Testing

Destructive tests have been developed to test the skill of the welder as well as to check the quality of welded joints for the various types of metal. Common tests are tensile, izod, hardness, bend, and nick break. Tensile and izod tests were described in Chapter2 and Chapter3. The destructive tests used on the Alaskan pipeline are shown in Fig. 15-1.

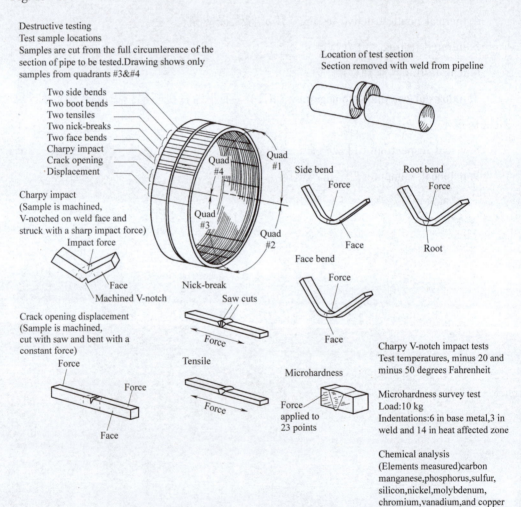

Fig. 15-1　The types of destructive tests that were used on the Alaskan pipeline

学习笔记

参考译文

破坏性测试

破坏性测试主要用来测试焊工的技能以及检查各种类型金属的焊接接头质量。常见的破坏性测试有拉伸试验、悬臂冲击试验、硬度试验、弯曲试验和缺口断裂试验。在第 2 章和第 3 章中已经描述了拉伸试验和悬臂冲击试验。图 15-1 所示为阿拉斯加管道上使用的破坏性试验。

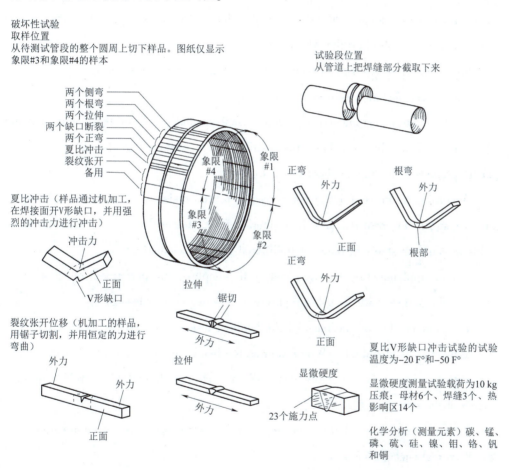

图 15-1　阿拉斯加管道上使用的破坏性试验类型

Reading Materials

Test Adhesion of Coating

Experiments were divided into two stages. In the first stage, samples were tested at five pieces without previous treatment of surface by Al coating by plasma jets before applying coating steel. Thickness originally deposited Al coating from supplier for all samples was measured by Thickness Elcometer 456 and measurement results are documented in the sample. Measurement of thickness was also applied to the coating steel which was applied on the Al coating.

Steel coating serves as a surface treatment of steel (heat resistant up to 600 ℃, the aesthetic effect, and increased corrosion resistance).

Steel coating was subjected by chemical analysis and was found that coating contain 13.6 wt% Si, which strengthens the coating and contributes to the deterioration of plastic properties in shaping sheet metal (plate).

According to the supplier of company Faren Ltd. resists such as acids, salt vapors and is resistant to corrosion and atmospheric effect. It can be used for adjustment of welds in stainless steels in place of traditional pickling methods.

Before applying the coating steel, all samples was performed on thoroughly degrease the surface with Al using highly effective degreaser Cleas-S.

It was found that on the basis of the bending tests in the test preparation adhesion of coating steel is unsuitable. When forming the bend into the shape of "U" in the test preparation occurred in the interval of radii of curvature from R11 to R35 mm steel breach of the coating (Fig. 15-2).

In the second stage of the experiment coating steel was used in conjunction with MU Brno multi-jet plasma system.

New Words

1. carbon　碳（化学元素）
2. manganese　锰
3. phosphorus　磷

Fig. 15-2　Photo of damaged coating steel after the bending without application of plasma

4. sulfur　硫

5. silicon　硅

6. nickel　镍

7. molybdenum　钼

8. chromium　铬

9. vanadium　钒

10. copper　铜

Phrases and Expressions

1. lamellar tearing　层状撕裂

2. hot crack　热裂纹

3. cold crack　冷裂纹

4. reheat crack　再热裂纹

5. stress corrosion crack（SCC）　应力腐蚀裂纹

Chapter 16 Quality Assurance

The quality of a product may be stated in terms of a measure of the degree to which it conforms to specifications and standards of workmanship. [①] These specifications and standards should reflect the degree to which the product satisfies the wants of a particular customer or user. The quality assurance function is charged with the responsibility of maintaining product quality consistent with those requirements and it involves the following four basic activities:

(1) Quality specification;

(2) Inspection;

(3) Quality analysis;

(4) Quality control (QC).

Quality Specification The product design engineer provides the basic specifications of product quality by means of various dimensions, tolerances, and other requirements cited on the engineering drawings. These basic specifications are further refined and elaborated upon on the manufacturing drawings which are used by the methods engineering personnel to specify the manufacturing and fabrication procedures. In many cases, the quality assurance group will provide a further interpretation of those specifications as a basis for specifying inspection procedures.

Inspection The mass production of interchangeable products is not altogether effective without some means of appraising and controlling product quality. Production operators are required, for example, to machine a given number of parts to the specifications shown on a drawing. There are often many factors that cause the parts to deviate from specifications during the various manufacturing operations. A few of these factors are variations in raw materials, deficiencies in machines and tools, poor methods, excessive production rates, and human errors.

Provisions must be made to detect errors so that the production of faulty parts can be stopped. The inspection department is usually charged with this responsibility, and its job is to interpret the specifications properly, inspect for conformance to those

specifications, and then convey the information obtained to the production people, who can make any necessary corrections to the process. Inspection operations are performed on raw materials and purchased parts at "receiving inspection". "In-process inspection" is performed on products during the various stages of their manufacture, and finished products may be subjected to "final inspection".

Quality Analysis During the various inspection processes a variety of quality information is recorded for review by representatives from both manufacturing and quality assurance. In essence, this information provides a basis for analyzing the quality of the product as it relates to the capability of the manufacturing process and to the quality specifications initially prescribed for the product. Generally, it is not possible to mass produce products that are 100% free of defects except at considerable expense. The quality analysis will determine the level of defectivity of the product so that a decision can be made whether or not that level is tolerable. ②If a given level of defectivity is not tolerable, decisions must be made relative to ①what to do with the process, and ②what to do with the product.

Quality Control The word "control" implies regulation, and of course, regulation implies observation and manipulation. Thus a pilot flying an aircraft from one city to another must first set it on the proper course heading and then must continue to observe the progress of the craft and manipulate the controls so as to maintain its flight path in the proper direction. Quality control in manufacture is an analogous situation. It is simply a means by which management can be assured that the quality of product manufactured is consistent with the quality-economy standards that have been established. The word "quality" does not necessarily mean "the best" when applied to manufactured products. It should imply "the best for the money".

Questions

1. What is meant by the term "interchangeable manufacture"?

2. Name and describe the four basic activities commonly involved in the quality assurance function.

3. Why is the inspection process necessary in most mass-production activities?

4. During what stages in the manufacturing process are inspection operations

commonly performed?

Notes

①The quality of a product may be stated in terms of a measure of the degree to which it conforms to specifications and standards of workmanship.

产品的质量可以用测量结果与产品的技术规格和标准的吻合程度来表达。

句中 in terms of 可译为"就……来说，根据，用……话来说"，而 to which it conforms to specifications and standards of workmanship 引导定语从句。

②The quality analysis will determine the level of defectivity of the product so that a decision can be made whether or not that level is tolerable.

质量分析决定着产品存在缺陷的程度，以便决定有这个缺陷存在的产品是否可以接受。

句子 so that a decision can be made whether or not that level is tolerable 是由 so that 引导的目的状语从句，可译为"为了，以便"；从句中 that 引导的是一个限定性定语从句。

参考译文

质量保证

产品的质量可以用测量结果与产品的技术规格和标准的吻合程度来表达。这些规格和标准应当体现产品对一个消费者或用户要求的满意程度。质量保证的作用是具有维护产品质量与要求相一致的职责，包括以下四个基本事项：

（1）质量规格；

（2）检验；

（3）质量分析；

（4）质量控制。

质量规格　产品设计是靠来自工程图纸上产品的各种尺寸、公差和其他一些要求为产品质量提供一个最基本的规定。这些规定在工程图中进行一些修改和细化，工程图是工程人员用来评述制造和装配工艺流程的。在多数情况下，质量控制组将提供关于技术规定更详细的解释，作为技术检验程序的基础。

检验　可互换产品的批量生产没有一些产品评估和质量控制的手段是不会

完全有效的。比如，生产操作过程中是需要按图纸上显示的技术规格制造给定数量的产品的，在各种制造工作中，经常有很多造成零件与技术规格相偏离的因素。其中少数因素是原材料、机器和工具的缺陷，方法落后、过高的生产率和人为错误等。

供应品必须进行缺陷检验，以保证阻止缺陷产品的生产，检验部门通常要担负这个责任。这个部门的工作就是正确解释技术规格、检验产品与规格的一致性，然后把这个信息传给生产者，使他们能在工艺中做必要的改正。针对原材料和购买的零件进行的检验叫作"接收检验"。"工艺检验"是在产品制造的不同阶段进行的；成品件的检验则为"最终检验"。

质量分析　在各种检验活动中，把种种质量信息记录下来以供生产厂家和质量保证部门的双方代表来复检。因为这种信息是与生产能力和起初给产品所规定的质量技术规定等相联系，所以它在本质上给分析产品的质量提供了基础。通常，批量生产的产品除了特别昂贵的以外，其他的达到100%的无缺陷是不可能的。质量分析对产品的缺陷程度的评价起决定作用，以便决定这个程度是否允许。

如果给定的缺陷水平是不可接受的，则必须就以下方面做出决定：①如何处理流程，以及②如何处理产品。

质量控制　"控制"这一词隐含着调整。当然，调整就意味着检查和处理。因此，一位飞行员驾驶一架飞机从一个城市飞往另一个城市，他首先在一个正确的航标的指引下飞行，然后必须不断地观察各级的进程和操作控制器，以保证正确的航线。生产者的质量控制与飞行的作用相似。通过管理部门来保证生产产品的质量与已建立的质量-经济标准相一致，这仅是一种手段。当把质量应用到制造产品时，"质量"一词不必要意味着最好，它仅意味着"货真价实"。

 Reading Material

Statistical Quality Control

Increasing demands for higher quality of product often result in increased cost of inspection and surveillance of processes and product. The old philosophy of attempting to inspect quality into a product is time consuming and costly. It is, of course, more sensible to control the process and make the product correctly rather than rely on having to sort out defective product from good product. Certain statistical techniques have been developed which provide economical means of maintaining continual analysis and control of processes and product.

These are known as statistical quality control methods, and many of them are formulated on the following basic statistical concepts:

(1) A population or universe is the complete collection of objects or measurements of the type in which we are interested at a particular time. The population may be finite or infinite.

(2) A sample is a finite group or set of objects taken from a population.

(3) The average is a point or value about which a population or a sample set of measurements tends to cluster. It is a measure of the ordinariness or central tendency of a group of measurements.

(4) Variation is the tendency for the measurements or observations in a population or a sample to scatter or disperse themselves about the average value.

In many cases the sizes observed from the inspection of a dimension of a group of pieces have been found to be distributed as shown in the figure below. If the pins represented include all the existing pins of a particular type, then this would be referred to as a population distribution. If those pins represent only a portion of a larger batch of the same type pins, then it would be called a sample distribution.

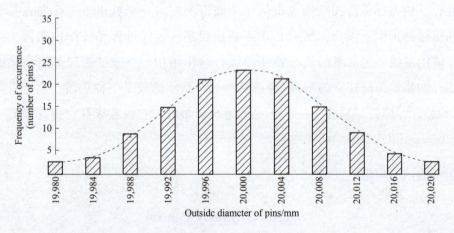

Frequency distribution of outside diameters of ground pins

About "Defect" & "Flaw"

The term "flaw" and "defect" have been used interchangeably and neither has been taken to signify either an acceptable or an unacceptable condition. More neutral

terms such as "discontinuity" "imperfection", or "inhomogeneity" are too cumbersome for general use, and the term "flaw sensitivity" and "defect delectability" are so widespread in NDT, and have been used for so many years, that it seems unnecessary to propose anything different. It is understood however, that for legal purposes the EEC has ruled that the term "defect" signifies that the material, fabrication, etc. is defective i. e. unserviceable. The term "flaw" should therefore be used for any imperfections which are not considered to be jectable. Thus, on this interpretation of the words, there is no such thing as an "acceptable detect".

Quality Control and NDT

This is known variously as nondestructive testing, nondestructive evaluation, nondestructive characterization, or nondestructive inspection. Quality control, quality technology, and noncontact measurements are related subjects that include or use NDT techniques.

There is concern for the most minute detail that may affect the future performance of the object in service, so that all properties need to be under control and all factors understood that may lead to breakdown.

The established test methods include radiography, ultrasonics testing, magnetic particle inspection, liquid penetrant testing (PT), thermography, electrical and magnetic methods, and visual-optical testing (VT).

In the case of radiography, X-ray and γ-ray are well established, but neutron, proton, and Compton scattering are also used and there have been recent important advances in tomography.

Because NDT is all-encompassing, it is most useful to have a library of as many examples or case studies as possible, all concerned with practical situations.

New Words

1. workmanship   *n.* 手艺，技艺，技巧；工艺水平
2. responsibility   *n.* 责任，义务，职责
3. cite   *vt.* 引用，引证
4. elaborate   *adj.* 精心制作的，详细阐述的；*v.* 详细阐述

5. interpretation *n.* 解释，说明；分析，整理

6. interchangeable *adj.* 可互换的，可交换的

7. appraise *v.* 评估，评价，鉴定

8. deviate *vi.* 背离；偏离；违背

9. deficiency *n.* 缺乏，不足，缺陷

10. faulty *adj.* 不合格的，没用的，报废的

11. conformance *n.* 顺应，一致

12. purchase *vt.* 买，购买

13. representative *n.* 代表；*adj.* 典型的

14. prescribe *v.* 指示，建议；*n.* 指标，规定

15. manipulation *n.* 处理，操作，操纵，应付

16. analogous *adj.* 类似的，相似的

Glossary of Terms

1. quality assurance 质量保证

2. quality specification 质量说明书

3. engineering drawing 工程图

4. mass production 成批生产，大批生产

5. raw material 原材料

6. finished product 成品

7. quality analysis 质量分析

8. quality control 质量控制

Glossary of Terms for Quality Control

1. quality 质量

Degree to which a set of inherent characteristics fulfils requirements.

2. quality control 质量控制

Part of quality management focused on fulfilling quality requirements.

3. quality assurance 质量保证

Part of quality management focused on providing confidence that quality requirements will be fulfilled.

4. inspection　检验

Conformity evaluation by observation and judgment accompanied as appropriate by measurement, testing or gauging.

5. test　试验

Determination of one or more characteristics.

6. verification　验证

Confirmation, through the provision of objective evidence, that specified requirements have been fulfilled.

7. conformity　合格

Fulfillment of a requirement.

8. nonconformity　不合格

Non-fulfillment of a requirement.

9. defect　缺陷

Any discontinuity that interferes with the usefulness or service of a part.

10. flaw　缺陷

A discontinuity in a material that prohibits the use of that material for a specific purpose.

11. discontinuity　不连续

Any interruption in the physical configuration or composition of a part. It may or may not be a defect.

12. evaluation　评价

Determination of whether an indication will be detrimental to the service of a part.

13. rework　返工

Action on a nonconforming product to make it conform to the requirements.

14. regrade　降级

Alteration of the grade of a nonconforming product in order to make it conform to requirements differing from the initial ones.

15. repair　返修

Action on a nonconforming product to make it acceptable for the intended use.

（repair includes remedial action taken on a previously conforming product to restore it for use, for example as part of maintenance）.

16. scrap　报废

Action on a nonconforming product to preclude its originally intended use（example：recycling, destruction）.

17. concession　让步

Permission to use or release a product that does not conform to specified requirements.

18. reliability　可靠性

Reliability is the extent to which an experiment, test, or any measuring procedure yields the same result on repeated trials.

Review Questions

Based on the professional knowledge learned，complete the following exercises.

一、Translation from English to Chinese（英译汉）

1. tension （   ）

2. modulus （   ）

3. elongation （   ）

4. ductility （   ）

5. impression （   ）

6. notch （   ）

7. fatigue （   ）

8. ultrasonic （   ）

9. radiography （   ）

10. frequency （   ）

11. component （   ）

12. transducer （   ）

13. attenuate （   ）

14. amplitude （   ）

15. couple （   ）

二、Translation from Chinese to English（汉译英）

1. 无损检测 （   ）

2. 夹杂物 （   ）

3. 液体渗透法 （   ）

4. 超声信号 （   ）

5. 发射探头 （   ）

6. 工业用射线 （   ）

7. γ射线 （   ）

8. 导电材料（   ）

9. 涡流（   ）

10. 磁粉探伤（   ）

11. 冲击载荷（   ）

12. 布氏硬度（   ）

三、Translate the following sentences into Chinese（根据专业知识，翻译以下句子）

1. When a material under tension reaches the limit of its elastic strain and begins to flow plastically，it is said to have yielded.

T：_____

_____

2. To determine the elongation，the increase in distance between two reference marks，scribed on the workpiece before test，is measured with the two halves of the broken workpiece held together.

T：_____

_____

3. Several different types of testing machines have been constructed in which the stress is applied by bending，torsion，tension or compression，but all involve the same principle of subjecting the material to constant cycles of stress.

T：_____

_____

4. Compression mounting，the most common mounting method，uses pressure and heat to encapsulate the workpiece with a thermosetting or thermoplastic mounting material.

T：_____

_____

5. By the proper application of heat，as in annealing，these work-hardened metals can be made to recrystallize，making them softer，more ductile and more amenable to further manufacturing processes.

T：_____

_____

6. The signals are sent through the part and the time intervals that elapse between the initial pulse and the arrival of the various echoes are displayed on an

oscilloscopescreen.

T：_____

_____

四、Based on professional knowledge，answer the following questions in English（根据专业知识，用英语回答以下问题）

1. Why were wood and stone supplanted by metals?

2. What can be calculated from the long-elongation or stress-strain curve?

3. What is the tensile strength?

4. How to determine the elongation?

5. Is a material which may have a high tensile strength suitable for shock loading conditions? If so，why?

6. What is the hardness of a material?

7. What is a liquidus line? What is a solidus line?

8. What are the three three-phase reactions that appear in the iron-carbon phase diagram?

9. Describe the procedure for preparing a metallographic workpiece.

10. Why should the sample be mounted?

11. When grinding，what should you attention to?

12. What types of materials may be heat treated? Why?

13. What is the NDT?

14. What is the visual inspection? Describe its uses.

15. What is the weldability of metals?

16. What happens when welding stainless steel?

17. How many factors are there that influence the weldability of metals?

18. Describe the applications of the ultrasonic inspection.

19. What is the advancement of the immersion inspection?

20. Presently，what types of penetrating are used for industrial radiography?

21. How are the X-rays generated?

22. What differences have the generation of X-rays with γ-rays and neutron beams?

23. What is the principle of magnetic particle testing?

24. What is the difference between the residual technique and the continuous technique?

25. What is the principle of liquid penetrant testing?

26. What are the basic steps of liquid penetration testing?

27. How do eddy-current tests differ from magnetic-particle testing?

28. What are the advantages of TOFD?

29. Why is the inspection process necessary in most mass-production activities?

30. During what stages in the manufacturing process are inspection operations commonly performed?

Appendix Nondestructive Testing Terms in Chinese and English 中英文无损检测名词术语

A

absorbed dose　吸收剂量

absorbed dose rate　吸收剂量率

acceptance limits　验收范围

acceptance level　验收水平

acceptance standard　验收标准

accumulation test　累积检测

AC magnetic permeability　交流磁导率

AC magnetic saturation　交流磁饱和

acoustic emission count（emission count）　声发射计数（发射计数）

acoustic emission transducer　声发射换能器（声发射传感器）

acoustic emission（AE）　声发射（AE）

acoustic holography　声全息术

acoustic impedance　声阻抗

acoustic impedance matching　声阻抗匹配

acoustic impedance method　声阻法

acoustic wave　声波

acoustical lens　声透镜

acoustic-ultrasonic（AU）　声–超声（AU）

activation　活化

activity　活度

adequate shielding　安全屏蔽

aguenons developer　水性显像剂

ampere turn　安匝数

amplitude　幅度

angle beam method　斜射法

angle of incidence　入射角

angle of reflection　反射角

angle of spread　指向角

angle of squint　偏向角

angle probe　斜探头

angstrom unit　埃（Å）

area amplitude response curve　面积幅度曲线

area of interest　评定区

artifical discontinuity　人工不连续性

artificial defect　人工缺陷

artificial discontinuity　标准人工缺陷

A-scan；A-scope　A 型扫描

attenuation coefficient　衰减系数

attenuator　衰减器

audible leak indicator　音响泄漏指示器

automatic testing　自动检测

autoradiography　自动射线照相

avaluation　评定

aguenons developer　水性显像剂

B

barium concrete　钡混凝土

barn　靶

base fog　片基灰雾

bath　槽液

Bayard-Alpert ionization gage　B-A 型电离计

beam　光/声束

beam ratio　束比

beam angle　束角

beam axis　束轴线

beam index　束入射点

beam path location　束程定位

beam path；path length　束程

beam spread　束扩散

betatron　电子感应加速器

bimetallic strip gage　双金属片计

bipolar field　双极磁场

black light filter　黑光滤波器

black light　黑光

blackbody　黑体

blackbody equivalent temperature　黑体等效温度

Bleakney mass spectrometer　波利克尼质谱仪

bleedout　渗出

bottom echo　底面回波

bottom surface　底面

boundary echo（first）　边界一次回波

bremsstrahlung　韧致辐射

broad-beam condition　宽射束条件

brush application 刷涂

B-scan presentation　B 型扫描显示

B-scope；B-scan　B 型扫描

C

calibration instrument　设备校准

capillary action　毛细管作用

carrier fluid　载液

carry over of penetrate　渗透剂移转

cassette　暗盒

cathode　阴极

central conductor　中心导体

central conductor method　中心导体法

characteristic curve　特性曲线

characteristic curve of film　胶片特性曲线

characteristic radiation　特征辐射

chemical fog　化学灰雾

cine-radiography　射线（活动）电影摄影术

contact pads　接触垫

circumferential coil　圆环线圈

circumferential field　周向磁场

circumferential magnetization method　周向磁化法

clean　清理

clean-up　清除

clearing time　定透时间

coercive force　矫顽力

coherence　相干性

coherence length　相干长度（谐波列长度）

coil test　测试线圈

coil size　线圈大小

coil spacing　线圈间距

coil technique　线圈技术

coil method　线圈法

coincidence discrimination　符合鉴别

cold-cathode ionization gage　冷阴极电离计

collimator　准直器

collimation　准直

combined colour contrast and fluorescent penetrant　着色荧光渗透剂

compressed air drying　压缩空气干燥

compressional wave　压缩波

compton scatter　康普顿散射

continuous emission　连续发射

continuous linear array　连续线阵

continuous method　连续法

continuous spectrum　连续谱

continuous wave　连续波

contrast stretch　对比度宽限

contrast　对比度

contrast agent　对比剂

contrast aid　反差剂

contrast sensitivity　对比灵敏度

control echo　监视回波

couplant　耦合剂

coupling　耦合

coupling loss　耦合损失

cracking　裂解

creeping wave　爬波

critical angle　临界角

cross section　横截面

cross talk　串音

cross drilled hole　横孔

crystal　晶片

C-scope；C-scan　C 型扫描

Curie point　居里点

Curie temperature　居里温度

Curie（Ci）　居里（Ci）

current flow method　通电法

current induction method　电流感应法

current magnetization method　电流磁化法

cut-off level　截止电平

D

DC magnetic permeability　直流磁导率

dead zone　盲区

decay curve　衰变曲线

decibel（dB）　分贝（dB）

defect　缺陷

defect detection sensitivity　缺陷检出灵敏度

defect resolution　缺陷分辨力

definition　清晰度

demagnetization　退磁

demagnetization factor　退磁因子

demagnetizer　退磁装置

densitometer　黑度计

density　黑度（底片）

density comparison strip　黑度比较片

detecting medium　检验介质

detergent remover　清洗剂

developer　显像剂

developer station　显像工位

nonaqueous developer（sus-pendible）　非水（可悬浮）显像剂

developing time　显像时间

development　显像

diffraction mottle　衍射斑

diffuse indication　松散指示

diffusion　扩散

digital image acquisition system　数字图像识别系统

dilatational wave　膨胀波

dip and drain station　浸渍和流滴工位

direct contact magnetization　直接接触磁化

direct exposure imaging　直接曝光成像

direct contact method　直接接触法

directivity　指向性

discontinuity　不连续性

distance-gain-size　距离-增益-尺寸

distance marker　距离刻度

dose equivalent　剂量当量

dose rate meter　剂量率计

dosemeter　剂量计

double crystal probe　双晶片探头

double probe technique　双探头法

double transceiver technique　双发双收法

double traverse technique　二次波法

dragout　带出

drain time　滴落时间

drift　漂移

dry method　干法

dry powder　干粉

dry technique　干粉技术

dry developer　干显像剂

dry developing cabinet　干显像柜

dry method　干粉法

drying oven　干燥箱

drying station　干燥工位

drying time　干燥时间

D-scope；D-scan　D 型扫描

dual search unit　双探头

dual-focus tube　双焦点管

duplex-wire image quality indicator　双线像质指示器

duration　持续时间

dwell time　停留时间

dye penetrant　着色渗透剂

dynamic leak test　动态泄漏检测

dynamic leakage measurement　动态泄漏测量

dynamic range　动态范围

dynamic radiography　动态射线透照术

E

echo　回波

echo frequency　回波频率

echo height　回波高度

echo indication　回波指示

echo transmittance of sound pressure　往复透过率

echo width　回波宽度

eddy current　涡流

eddy current flaw detector　涡流探伤仪

eddy current testing　涡流检测

edge　端面

edge echo　棱边回波

edge effect　边缘效应

effective depth penetration（EDP）　有效穿透深度（EDP）

effective focus size　有效焦点尺寸

effective magnetic permeability　有效磁导率

effective permeability　有效磁导率

effective reflection surface of flaw　缺陷有效反射面

effective resistance　有效电阻

elastic medium　弹性介质

electric displacement　电位移

electrical center　电中心

electrode　电极

electromagnet　电磁铁

electro magnetic acoustic transducer　电磁声换能器

electromagnetic induction　电磁感应

electromagnetic radiation　电磁辐射

electromagnetic testing　电磁检测

electro-mechanical coupling factor　机电耦合系数

electron pair production　电子对生成

electron radiography 电子射线照相

electron volt 电子伏特

electronic noise 电子噪声

electrostatic spraying 静电喷涂

emulsification 乳化

emulsification time 乳化时间

emulsifier 乳化剂

encircling coil 环绕式线圈

end effect 端部效应

energizing cycle 激励周期

equalizing filter 均衡滤波器

equivalent 当量

equivalent image quality indicator sensitivity 像质指示器当量灵敏度

equivalent nitrogen pressure 等效氮压

equivalent penetrameter sensitivity 透度计当量灵敏度

equivalent method 当量法

etching 浸蚀

evaluation 评定

evaluation threshold 评价阈值

event count 事件计数

event count rate 事件计数率

examination area 检查范围

examination region 检查区域

exhaust pressure/discharge pressure 排气压力

exhaust tubulation 排气管道

expanded time-base sweep 时基线展宽

exposure 曝光

exposure table 曝光表格

exposure chart 曝光曲线

exposure fog 曝光灰雾

exposure，radiographic exposure 曝光，射线照相曝光

extended source 扩展源

F

facility scattered neutrons　条件散射中子

false indication　假指示

family　族

far field　远场

feed-through coil　穿过式线圈

resultant magnetic field　复合磁场

fill factor　填充系数

film badge　胶片襟章剂量计

film base　片基

film contrast　胶片对比度

film gamma　胶片 γ 值

film processing　胶片冲洗加工

film speed　胶片感光度

film unsharpness　胶片不清晰度

film viewing screen　观察屏

filter　滤波器/滤光板

final test　复探

flat-bottomed hole　平底孔

flat-bottomed hole equivalent　平底孔当量

flaw　缺陷

flaw characterization　缺陷特性

flaw echo　缺陷回波

flexural wave　弯曲波

floating threshold　浮动阈值

fluorescence　荧光

fluorescent inspection method　荧光检验法

fluorescent magnetic particle inspection　荧光磁粉检验

fluorescent dry deposit penetrant　干沉积荧光渗透剂

fluorescent light　荧光

fluorescent magnetic powder　荧光磁粉

fluorescent penetrant　荧光渗透剂

fluorescent screen　荧光屏

fluoroscopy　荧光检查法

flux leakage field　磁通泄漏场

flux line　磁通线

focal spot　焦点

focal distance　焦距

focus length　焦点长度

focus size　焦点尺寸

focus width　焦点宽度

focus（electron）　电子焦点

focused beam　聚焦声束

focusing probe　聚焦探头

focus-to-film distance（f. f. d）　焦点-胶片距离（焦距）

fog　底片灰雾

fog density　灰雾密度

footcandle　英尺烛光

freguency　频率

frequency constant　频率常数

fringe　干涉带

front distance　前沿距离

front distance of flaw　缺陷前沿距离

full-wave direct current（FWDC）　全波直流（FWDC）

fundamental frequency　基频

furring　毛状痕迹

G

gage pressure　表压

gain　增益

gamma radiography　γ射线照相术

gamma ray source   γ射线源

gamma ray source container   γ射线源容器

gamma rays   γ射线

gamma-ray radiographic equipment   γ射线照相装置

gap scanning   间隙扫查

gas   气体

gate   闸门

gating technique   选通技术

Gauss   高斯

Geiger-Muller counter   盖革·弥勒计数器

geometric unsharpness   几何不清晰度

Gray（Gy）   戈瑞（Gy）

grazing incidence   掠入射

grazing angle   掠射角

group velocity   群速度

H

half life   半衰期

half-wave current（HW）   半波电流

half-value layer（HVL）   半值层（HVL）

half-value method   半波高度法

halogen   卤素

halogen leak detector   卤素检漏仪

hard X-ray   硬X射线

hard-faced probe   硬膜探头

harmonic analysis   谐波分析

harmonic distortion   谐波畸变

harmonics   谐频

head wave   头波

helium bombing   氦轰击法

helium drift   氦漂移

helium leak detector　氦检漏仪

hermetically tight seal　气密密封

high vacuum　高真空

high energy X-ray　高能 X 射线

optical holography　光全息照相

acoustic holography　声全息

hydrophilic emulsifier　亲水性乳化剂

hydrophilic remover　亲水性洗净剂

hydrostatic text　流体静力检测

hysteresis　磁滞

I

ID coil　ID 线圈

image definition　图像清晰度

image contrast　图像对比度

image enhancement　图像增强

image magnification　图像放大

image quality　图像质量

image quality indicator sensitivity　像质指示器灵敏度

image quality indicator（IQI）/image quality indication　像质指示器

imaging line scanner　图像线扫描器

immersion probe　液浸探头

immersion rinse　浸没清洗

immersion testing　液浸法

immersion time　浸没时间

impedance　阻抗

impedance plane diagram　阻抗平面图

imperfection　不完整性

impulse eddy current testing　脉冲涡流检测

incremental permeability　增量磁导率

indicated defect area　缺陷指示面积

indicated defect length　缺陷指示长度

indication　指示

indirect exposure　间接曝光

indirect magnetization　间接磁化

indirect magnetization method　间接磁化法

indirect scan　间接扫描

induced field　感应磁场

induced current method　感应电流法

infrared imaging system　红外成像系统

infrared sensing device　红外扫描器

inherent fluorescence　固有荧光

infrared thermography　红外热成像

inherent filtration　固有滤波

initial permeability　起始磁导率

initial pulse　始脉冲

initial pulse width　始波宽度

inserted coil　插入式线圈

inside coil　内部线圈

inside-out testing　外泄检测

inspection　检验

inspection medium　检验介质

inspection frequency/test frequency　检验/测频率

intensifying factor　增感系数

intensifying screen　增感屏

interface boundary　界面

interface echo　界面回波

interface trigger　界面触发

interference　干涉

interpretation　解释

ion pump　离子泵

ion source　离子源

ionization chamber　电离室

ionization potential　电离电位

ionization vacuum gage　电离真空计

ionography　电离射线照相术

irradiance　辐射通量密度

isolation　隔离检测

isotope　同位素

K

K value　*K* 值

Kaiser effect　凯塞效应

kilo volt（kV）　千伏特

kiloelectron volt（keV）　千电子伏特

krypton 85　氪 85

L

L/D ratio　*L/D* 比

Lamb wave　兰姆波

latent image　潜像

lateral scan　左右扫描

lateral scan with oblique angle　斜平行扫描

latitude（of an emulsion）　胶片宽容度

lead screen　铅屏

leak　泄漏孔

leak detector　检漏仪

leak testtion　泄漏检测

leakage field　泄漏磁场

leakage rate　泄漏率

leech　磁吸盘

lift-off effect　提离效应

light intensity　光强度

limiting resolution　极限分辨率

line scanner　线扫描器

line focus　线焦点

line pair pattern　线对检测图

line pairs per millimetre　每毫米线对数

linear（electron）accelerator（LINAC）　电子直线加速器（LINAC）

linear attenuation coefficient　线衰减系数

linear scan　线扫描

linearity（time or distance）　线性（时间或距离）

linearity，amplitude　幅度线性

lines of force　磁力线

lipophilic emulsifier　亲油性乳化剂

lipophilic remover　亲油性洗净剂

liquid penetrant examination　液体渗透检查

liquid film developer　液膜显像剂

local magnetization　局部磁化

local magnetization method　局部磁化法

local scan　局部扫描

localizing cone　定域喇叭筒

location　定位

location accuracy　定位精度

location computed　定位计算

location marker　定位标记

location upon ΔT　时差定位

location cluster　定位群集

location continuous AE signal　定位连续 AE 信号

longitudinal field　纵向磁场

longitudinal magnetization method　纵向磁化法

longitudinal resolution　纵向分辨率

longitudinal wave　纵波

longitudinal wave probe　纵波探头

longitudinal wave technique　纵波法

loss of back reflection　背面反射损失

love wave　乐甫波

low energy gamma radiation　低能 γ 辐射

low-energy photon radiation　低能光子辐射

luminance　亮度

luminosity　流明

lusec　流西克

M

maga or million electron volt MeV　兆电子伏特

magnetic history　磁化史

magnetic hysteresis　磁性滞后

magnetic particle field indication　磁粉磁场指示器

magnetic particle inspection flaw indications　磁粉检验的缺陷显示

magnetic circuit　磁路

magnetic domain　磁畴

magnetic field distribution　磁场分布

magnetic field indicator　磁场指示器

magnetic field meter　磁场计

magnetic field strength　磁场强度

magnetic field　磁场

magnetic flux　磁通

magnetic flux density　磁通密度

magnetic force　磁化力

magnetic leakage field　漏磁场

magnetic leakage flux　漏磁通

magnetic moment　磁矩

magnetic particle　磁粉

magnetic particle indication　磁痕

magnetic particle testing/magnetic particle examination　磁粉检测/查

magnetic permeability　磁导率

magnetic pole　磁极

magnetic saturation　磁饱和

magnetic storage meclium　磁储介质

magnetic writing　磁写

magnetizing　磁化

magnetizing current　磁化电流

magnetizing coil　磁化线圈

magnetostrictive effect　磁致伸缩效应

magnetostrictive transducer　磁致伸缩换能器

main beam　主声束

manual testing　手动检测

marker　时标

MA-scope；MA-scan　MA 型显示

masking　遮蔽

mass attenuation coefficient　质量衰减系数

mass number　质量数

mass spectrometer（MS）　质谱仪（MS）

mass spectrometer leak detector　质谱检漏仪

mass spectrum　质谱

master/slave discrimination　主/从鉴别

mean free path　平均自由程

medium vacuum　中真空

mega or million volt MV　兆伏特

micro focus X-ray tube　微焦点 X 光管

microfocus radiography　微焦点射线照相

micrometre　微米

micron of mercury　微米汞柱

microtron　电子回旋加速器

milliampere（mA）　毫安（mA）

millimetre of mercury　毫米汞柱

minifocus X-ray tube　小焦点调射线管

minimum detectable leakage rate　最小可探泄漏率

minimum resolvable temperature difference（MRTD） 最小可分辨温度差（MRTD）

mode 波型

mode conversion 波型转换

mode transformation 波型转换

moderator 慢化器

modulation transfer function（MTF） 调制转换功能（MTF）

modulation analysis 调制分析

molecular flow 分子流

molecular leak 分子泄漏

monitor 监视器

monochromatic 单色波

movement unsharpness 移动不清晰度

moving beam radiography 可动射束射线照相

multiaspect magnetization method 多向磁化法

multidirectional magnetization 多向磁化

multifrequency eddy current testing 多频涡流检测

multiple back reflections 多次背面反射

multiple reflections 多次反射

multiple back reflections 多次底面反射

multiple echo method 多次反射法

multiple probe technique 多探头法

multiple triangular array 多三角形阵列

N

narrow beam condition 窄射束

near field 近场

near field length 近场长度

near surface defect 近表面缺陷

net density 净（光学）密度

neutron 中子

neutron radiography 中子射线照相

newton（N） 牛顿

Nier mass spectrometer 尼尔质谱仪

noise 噪声

noise equivalent temperature difference（NETD） 噪声当量温度差（NETD）

nominal angle 标称角度

nominal frequency 标称频率

non-aqueous liquid developer 非水性液体显影剂

non-aqueous developer（suspendable） 非水（可悬浮）显像剂

noncondensable gas 非冷凝气体

nondestructivc examination（NDE） 无损检查（NDE）

nondestructive evaluation（NDE） 无损评价（NDE）

nondestructive inspection（NDI） 无损检验（NDI）

nondestructive testing（NDT） 无损检测（NDT）

non-erasable optical data 可固定光学数据

nonferromagnetic material 非铁磁性材料

nonrelevant indication 非相关指示

non-screen-type film 非增感型胶片

normal incidence 垂直入射（亦见直射声束）

normal permeability 标准磁导率

normal beam method; straight beam method 垂直法

normal probe 直探头

normalized reactance 归一化电抗

normalized resistance 归一化电阻

nuclear activity 核活性

nuclide 核素

O

object plane resolution 物体平面分辨率

object scattered neutron 物体散射中子

object beam 物体光束

object beam angle　物体光束角

object-film distance　被检体-胶片距离

over development　显影过度

over emulsfication　过乳化

overall magnetization　整体磁化

overload recovery time　过载恢复时间

overwashing　过洗

oxidation fog　氧化灰雾

P

palladium barrier leak detector　钯屏检漏仪

panoramic exposure　全景曝光

parallel scan　平行扫描

paramagnetic material　顺磁性材料

parasitic echo　干扰回波

partial pressure　分压

particle content　磁悬液浓度

particle velocity　质点（振动）速度

Pascal（Pa）　帕斯卡（帕）

Pascal cubic metres per second　帕立方米每秒

path length　光程长

path length difference　光程长度差

pattern　探伤图形

peak current　峰值电流

penetrameter　透度计

penetrameter sensitivity　透度计灵敏度

penetrant　渗透剂

penetrant comparator　渗透对比试块

penetrant flaw detection　渗透探伤

penetrant removal　渗透剂去除

penetrant station　渗透工位

penetration　穿透深度

penetration time　渗透时间

permanent magnet　永久磁铁

permeability coefficient　透气系数

phantom echo　幻象回波

phase analysis　相位分析

phase angle　相位角

phase controlled circuit breaker　断电相位控制器

phase detection　相位检测

phase hologram　相位全息

phase sensitive detector　相敏检波器

phase shift　相位移

phase velocity　相速度

phase-sensitive system　相敏系统

phillips ionization gage　菲利浦电离计

phosphor　荧光物质

photo fluorography　荧光照相术

photoelectric absorption　光电吸收

photographic emulsion　照相乳剂

photographic fog　照相灰雾

photostimulable luminescence　光敏发光

piezoelectric effect　压电效应

piezoelectric material　压电材料

piezoelectric stiffness constant　压电劲度常数

piezoelectric stress constant　压电应力常数

piezoelectric transducer　压电传感器

piezoelectric voltage constant　压电电压常数

Pirani gage　皮拉尼计

pitch and catch technique　一发一收法

pixel　像素

pixel size　像素尺寸

pixel disply size　像素显示尺寸

planar array　平面阵（列）

plane wave　平面波

plate wave　板波

plate wave technique　板波法

point source　点源

post emulsification　后乳化

post emulsifiable penetrant　后乳化渗透剂

post-cleaning　后清洗

powder　粉末

powder blower　喷粉器/喷枪

pre-cleaning　预清理

pressure difference　压力差

pressure dye test　压力着色检测

pressure probe　压力探头

pressure testing　压力检测

pressure-evacuation test　压力抽空检测

pressure mark　压痕

pressure design　设计压力

pre-test　初探

primary coil　一次线圈

primary radiation　初级辐射

probe gas　探头气体

probe test　探头检测

probe backing　探头背衬

probe coil　点/探头式线圈

probe coil clearance　探头线圈间隙

probe index　探头入射点

probe to weld distance　探头–焊缝距离

probe/search unit　探头

process control radiograph　工艺过程控制的射线照相

processing capacity　处理能力

processing speed　处理速度

prod　触头

projective radiography　投影射线照相

proportioning probe　比例探头

protective material　防护材料

proton radiography　质子射线照相

pulse　脉冲

pulse echo method　脉冲回波法

pulse repetition rate　脉冲重复率

pulse amplitude　脉冲幅度

pulse echo method　脉冲反射法

pulse energy　脉冲能量

pulse envelope　脉冲包络

pulse length　脉冲长度

pulse repetition frequency　脉冲重复频率

pulse tuning　脉冲调谐

pump-out tubulation　抽气管道

pump-down time　抽气时间

Q

Q factor　*Q* 值

quadruple traverse technique　四次波法

quality（of a beam of radiation）　（射线束的）质量

quality factor　品质因数

quenching　阻塞

quenching of fluorescence　荧光的猝灭

quick break　快速断间

R

Rad（rad）　拉德

radiance *L*　面辐射率 *L*

radiant existence *M*   辐射照度 *M*

radiant flux radiant power   辐射通量、辐射功率

radiation   辐射

radiation does   辐射剂量

radio frequency（RF）display   射频（RF）显示

radio-frequency mass spectrometer   射频质谱仪

radiograph   射线照片

radiographic contrast   射线照片对比度

radiographic equivalence factor   射线照相等效系数

radiographic exposure   射线照相曝光量

radiographic inspection   射线照相检测

radiographic quality   射线照相质量

radiographic sensitivity   射线照相灵敏度

radiographic contrast   射线底片对比度

radiographic equivalence factor   射线照相等效因子

radiographic inspection   射线照相检查

radiographic quality   射线照相质量

radiographic sensitivity   射线照相灵敏度

radiography   射线照相术

radiological examination   射线检查

radiology   射线学

radiometer   辐射计

radiometry   辐射测量术

radioscopy   射线检查法

range   量程

Rayleigh wave   瑞利波

Rayleigh scattering   瑞利散射

real image   实时图像

real-time radioscopy   实时射线检查法

rearm delay time   重新准备延时时间

reciprocity failure   倒易律失效

reciprocity law   倒易律

recording medium　记录介质

recovery time　恢复时间

rectified alternating current　脉动直流电

reference block　参考试块

reference beam　参考光束

reference block method　对比试块法

reference coil　参考线圈

reference line method　基准线法

reference standard　参考标准

reflection　反射

reflection coefficient　反射系数

reflection density　反射密度

reflector　反射体

refraction　折射

refractive index　折射率

reference beam angle　参考光束角

reicnlbation　网纹

reject；suppression　抑制

rejection level　拒收水平

relative permeability　相对磁导率

relevant indication　相关指示

reluctance　磁阻

Rem（rem）雷姆

remote controlled testing　机械化检测

replenisher　补充剂

representative quality indicator　代表性质量指示器

residual magnetic field/field，residual magnetic　剩磁场

residual technique　剩磁技术

residual magnetic method　剩磁法

residual magnetism　剩磁

resistance（to flow）气阻

resolution　分辨力

resonance method　共振法

response factor　响应系数

response time　响应时间

resultant field　复合磁场

resultant magnetic field　合成磁场

resultant magnetization method　组合磁化法

retentivity　保磁性

reversal　反转作用

ring-down count　振铃计数

ring-down count rate　振铃计数率

rinse　清洗

rise time　上升时间

rise-time discrimination　上升时间鉴别

rod-anode tube　棒阳极管

Roentgen　伦琴

roof angle　屋顶角

rotational magnetic field　旋转磁场

rotational magnetic field method　旋转磁场法

rotational scan　转动扫描

roughing　低真空

roughing line　低真空管道

roughing pump　低真空泵

S

safelight　安全灯

sampling probe　取样探头

saturation　饱和

magnetic saturation　磁饱和

saturation level　饱和电平

scan on grid line　格子线扫描

scan pitch　扫描间距

scanning　扫描

scanning index　扫描标记

scanning directly on the weld　焊缝上扫查

scanning path　扫描轨迹

scanning sensitivity　扫描灵敏度

scanning speed　扫描速度

scanning zone　扫描区域

scattered energy　散射能量

scatter unsharpness　散射不清晰度

scattered neutron　散射中子

scattered radiation　散射辐射

scattering　散射

Schlieren system　施利伦系统

scintillation counter　闪烁计数器

scintillator and scintillating crystals　闪烁器和闪烁晶体

screen　屏

screen unsharpness　荧光增感屏不清晰度

screen-type film　荧光增感型胶片

SE probe　SE 探头

search-gas　探测气体

second critical angle　第二临界角

secondary radiation　二次射线

secondary coil　二次线圈

secondary radiation　次级辐射

selectivity　选择性

semi-conductor detector　半导体探测器

sensitivity value　灵敏度值

sensitivity　灵敏度

sensitivity of leak test　泄漏检测灵敏度

sensitivity control　灵敏度控制

shear wave　切变波

shear wave probe　横波探头

shear wave technique　横波法

shim　薄垫片

shot　冲击通电

side lobe　副瓣

side wall　侧面

sievert（Sv）　希（Sv）

signal　信号

signal gradient　信号梯度

signal over load point　信号过载点

signal overload level　信号过载电平

signal to noise ratio　信噪比

single crystal probe　单晶片探头

single probe technique　单探头法

single traverse technique　一次波法

sizing technique　定量法

skin depth　集肤深度

skin effect　集肤效应

skip distance　跨距

skip point　跨距点

sky shine（air scatter）　空中散射效应

sniffing probe　嗅吸探头

soft X-ray　软 X 射线

soft-faced probe　软膜探头

solarization　负感作用

solenoid　螺线管

soluble developer　可溶显像剂

solvent remover　溶剂去除剂

solvent cleaner　溶剂清除剂

solvent removal penetrant　溶剂去除型渗透剂

sorption　吸着

sound diffraction　声绕射

sound insulating layer　隔声层

sound intensity 声强

sound intensity level 声强级

sound pressure 声压

sound scattering 声散射

sound transparent layer 透声层

sound velocity 声速

source 源

source data label 放射源数据标签

source location 源定位

source size 源尺寸

source-film distance 射线源-胶片距离

spacial frequency 空间频率

spark coil leak detector 电火花线圈检漏仪

specific activity 放射性比度

specified sensitivity 规定灵敏度

standard 标准

standard 标准工件

standard leak rate 标准泄漏率

standard leak 标准泄漏孔

standard test block 标准试块

standardization instrument 设备标准化

standing wave；stationary wave 驻波

step wedge 阶梯楔块

step-wedge calibration film 阶梯-楔块校准片

stereo-radiography 立体射线透照相术

subject contrast 被检体对比度

subsurface discontinuity 近表面不连续性

suppression 抑制

surface echo 表面回波

surface field 表面磁场

surface noise 表面噪声

surface wave 表面波

surface wave probe　表面波探头

surface wave technique　表面波法

surge magnetization　脉动磁化

surplus sensitivity　灵敏度余量

suspension　磁悬液

sweep　扫描

sweep range　扫描范围

sweep speed　扫描速度

swept gain　扫描增益

swivel scan　环绕扫查

system examination threshold　系统检验阈值

system inclacel artifact　系统感生物

system noise　系统噪声

T

tackground target　目标，本底

tandem scan　串列扫查

target　靶

television fluoroscopy　电视 X 射线荧光检查

temperature envelope　温度范围

tenth-value-layer（TVL）　十分之一值层（TVL）

test coil　检测线圈

test quality level　检测质量水平

test ring　试环

test block　试块

test frequency　试验频率

test piece　试片

test range　探测范围

test surface　探测面

thermal neutrons　热中子

thermocouple gage　热电偶计

thermogram   热谱图

thermoluminescent dosemeter（TLD）   热释光剂量计（TLD）

thickness sensitivity   厚度灵敏度

third critical angle   第三临界角

thixotropic penetrant   摇溶渗透剂

thormal resolution   热分辨率

threading bar   穿棒

three way sort   三档分选

threshold setting   阈值设置

threshold fog   阈值灰雾

threshold level   阈值水平

throttling   节流

transmission technique   穿透技术

through penetration technique   贯穿渗透法

through-coil technique   穿过式线圈技术

throughput   通气量

tight   密封

total reflection   全反射

total image unsharpness   总的图像不清晰度

tracer probe leak location   示踪探头泄漏定位

tracer gas   示踪气体

transducer   换能器/传感器

transition flow   过渡流

translucent base media   半透明载体介质

transmission   透射

transmission densitometer   透射密度计

transmission coefficient   透射系数

transmission point   透射点

transmission technique   透射技术

transmittance   透射率

transmitted film density   检测底片黑度

transmitted pulse　发射脉冲

transverse resolution　横向分辨率

transverse wave　横波

traveling echo　游动回波

travering scan；depth scan　前后扫查

triangular array　正三角形阵列

trigger/alarm condition　触发/报警状态

trigger/alarm level　触发/报警标准

triple traverse technique　三次波法

true continuous technique　准确连续法技术

true attenuation　真实衰减

tube current　管电流

tube head　管头

tube shield　管罩

tube shutter　管子光闸

tube window　管窗

tube-shift radiography　管子移位射线照相

two-way sort　两档分选

U

ultra-high vacuum　超高真空

ultrasonic leak detector　超声波检漏仪

ultrasonic testing　超声检测

ultrasonic noise level　超声噪声电平

ultrasonic cleaning　超声波清洗

ultrasonic field　超声场

ultrasonic flaw detection　超声探伤

ultrasonic flaw detector　超声探伤仪

ultrasonic microscope　超声显微镜

ultrasonic spectroscopy　超声频谱

ultrasonic testing system　超声检测系统

ultrasonic thickness gauge　超声测厚仪

ultraviolet radiation　紫外线辐射

under development　显影不足

unsharpness　不清晰

useful density range　有效光学密度范围

UV-A　A 类紫外线辐射

UV-A filter　A 类紫外线辐射滤片

V

vacuum　真空

vacuum cassette　真空暗盒

vacuum testing　真空检测

Van de Graaff generator　范德格拉夫起电机

vapor pressure　蒸汽压

vapour degreasing　蒸汽除油

variable angle probe　可变角探头

V path　V 型行程

vehicle　载体

vertical linearity　垂直线性

vertical location　垂直定位

visible light　可见光

vitual image　虚假图像

voltage threshold　电压阈值

W

wash station　水洗工位

water break test　水膜破坏试验

water column coupling method　水柱耦合法

water column probe　水柱耦合探头

water path；water distance　水程

water tolerance　水容限

water washable penetrant　可水洗型渗透剂

wave　波

wave guide acoustic emission　声发射波导杆

wave train　波列

wave form　波形

wave front　波前

wave length　波长

wave node　波节

wave train　波列

wedge　楔子

wet slurry technique　湿软磁膏技术

wet technique　湿法技术

wet method　湿粉法

wetting action　润湿作用

wetting action　润湿作用

wetting agent　润湿剂

wheel type probe；wheel search unit　轮式探头

white light　白光

white X-ray　连续 X 射线

wobble　摆动

wobble effect　抖动效应

working sensitivity　探伤灵敏度

wrap around　残响波干扰

X

xero radiography　静电射线照相

X-radiation　X 射线

X-ray controller　X 射线控制器

X-ray detection apparatus　X 射线探伤装置

X-ray film　射线胶片

X-ray paper　X 射线感光纸

X-ray tube　X 射线管

X-ray tube diaphragm　X 射线管光阑

Y

yoke　磁轭

yoke magnetization method　磁轭磁化法

Z

zigzag scan　锯齿扫查

zone calibration location　时差区域校准定位

zone location　区域定位

References

［1］KALPAKJIANS, SCHMIDS. 制造工程与技术［M］. 5 版. 北京：清华大学出版社，2006.

［2］RAO P N. 制造技术第 1 卷　铸造、成形和焊接［M］. 2 版. 北京：机械工业出版社，2006.

［3］SHULL P J. Nondestructive evaluation：theory, techniques, and applications ［M］. Boca Raton：CRC Press，2002.

［4］HELLIER C J. Handbook of nondestructive evaluation［M］. New York：McGraw-Hill Professional，2001.

［5］RAJ B，JAYAKUMAR T，THAVASIMUTHU M. Practical nondestructive testing ［M］. Cambridge：Wood head Publishing，2002.

［6］全燕鸣. 制造工程与技术（热加工）（英文版）及学习辅导（上册）［M］. 北京：机械工业出版社，2004.

［7］《新英汉机械工程词汇》编订组. 新英汉机械工程词汇［M］. 北京：科学出版社，2010.

［8］张培基，喻云根，李宗杰，等. 英汉翻译教程［M］. 2 版. 上海：上海外语教育出版社，2009.

［9］GOLIS M J. ASNT level Ⅲ study guide, ultrasonic testing method ［M］. 2nd ed. Ohio：The American Society for Nondestructive Testing, Inc，2008.

［10］KINSELLA T. ASNT level Ⅲ study guide, radiographic testing method ［M］. 2nd ed. Ohio：The American Society for Nondestructive Testing, Inc，2009.

［11］FENTON J D. ASNT level Ⅲ study guide, magnetic particle testing method ［M］. 2nd ed. Ohio：The American Society for Nondestructive Testing, Inc，2008.

［12］EICK C W. ASNT level Ⅲ study guide, liquid penetrant method ［M］. 2nd ed. Ohio：The American Society for Nondestructive Testing, Inc，2009.

［13］张小海，金信鸿. 无损检测专业英语［M］. 北京：机械工业出版社，2012.

［14］晏荣明. 无损检测专业英语［M］. 北京：机械工业出版社，2015.

学习笔记

[15] 洪宇翔. 焊接专业英语 [M]. 北京：机械工业出版社，2013.

[16] 王晓江. 热加工专业英语 [M]. 北京：机械工业出版社，2015.